Spiritual INTELLIGENCE

Activating the 4 Circuits of the Awakened Brain

DAWSON CHURCH

Energy Psychology Press
PO Box 222, Petaluma, CA 94953-0222
www.energypsychologypress.com

Church, Dawson.
Spiritual intelligence : activating the 4 circuits of the awakened brain / by Dawson Church — 1st ed.
p. cm.
ISBN: 978-1-60415-293-7 (Limited Gift First Edition)
ISBN: 978-1-60415-294-4 (Hardcover)
ISBN: 978-1-60415-295-1 (Softcover)
ISBN: 978-1-60415-296-8 (Epub)
ISBN: 978-1-60415-297-5 (Audio)
LC Subjects: Spirituality. | Awareness. | Insight. | Enlightenment—Miscellanea. | Neurosciences—Religious aspects.
LC Classification BL624 .N485 2025

Copyright © 2024, 2025 Dawson Church

The information in this book is not to be used to treat or diagnose any particular disease or any particular patient. Neither the authors nor the publisher is engaged in rendering professional advice or services to the individual reader. The ideas, procedures, and suggestions in this book are not intended as a substitute for consultation with a professional health care provider. Neither the authors nor the publisher shall be liable or responsible for any loss or damage arising from any information or suggestion in this book.

All rights reserved. No part of this publication may be reproduced, stored in a retrieval system, or transmitted in any form or by any means, electronic, mechanical, photocopy, recording, or otherwise, without prior written permission from Energy Psychology Press, with the exception of short excerpts used with acknowledgment of publisher and author.

Cover design by Victoria Valentine
Typesetting by Karin Kinsey
Editing by Stephanie Marohn
Typeset in ITC Galliard and Scriptina
Printed in China by China Best Printing
First Edition

10 9 8 7 6 5 4 3 2 1

Contents

Foreword

by Jean Houston

It is with profound admiration and anticipation that I introduce *Spiritual Intelligence,* a book that stands as both a beacon and a blueprint for humanity's next evolutionary leap. Dawson Church, a visionary, mystic, and scholar, leads us into realms of possibility in this work—a culmination of science and spirit, meticulously crafted and deeply transformative. In *Spiritual Intelligence,* Dawson takes us on an unprecedented journey into the untapped realms of our brains, minds, and consciousness, providing a new way to experience ourselves as *Homo spiritualis,* the possible human I have long described.

Throughout my life, I have studied and written about the human potential for self-evolution, exploring the undiscovered depths of our mind and spirit. My book *The Possible Human,* a work that continues to resonate for seekers today, delves into what I call "entelechy"—the innate drive within each of us, a formative principle encoded in our very cells—that propels us toward self-actualization.

Dawson Church's *Spiritual Intelligence* reveals that this drive is not merely poetic or metaphorical but a tangible physiological phenomenon. He identifies specific neural pathways—the "Enlightenment Network"—that lie dormant in our brains, waiting to be activated. These circuits, when engaged, open us to levels of compassion, creativity, and wisdom once accessible only to spiritual adepts. Dawson's work shows that anyone can access these heights of wisdom through practices designed to activate the brain circuits of a transcendent inner landscape.

On a global scale, Dawson's work resonates with my concept of "jump time," a term capturing the unprecedented opportunity humanity now faces. We are at a pivotal moment in history, a "jump time" characterized by rapid transformation. *Spiritual Intelligence* speaks to this moment not only as a philosophical concept but as an actual, physiological potential within each of us. Neuroscience reveals that our brains can adapt and transform, allowing us to rewire our neural pathways to embody our highest qualities. By engaging in these practices, we collectively shape the arc of our evolution, each of us a participant in this leap into a new era.

In *Spiritual Intelligence,* Dawson also explores the concept of the *social artist*—we, the architects of the future, who create and expand the boundaries of possibility. Those glimpsing these evolutionary insights are called to bring them into our culture, integrating them into all our social systems. Dawson envisions a world in which our growth in compassion leads to systemic changes—from education reform to holistic healthcare, from conflict resolution to ecological stewardship. This thrilling vision places *Spiritual Intelligence* at the forefront of our time, serving as both a personal guide and a manifesto for a future we consciously co-create.

At the core of Dawson's work is the insight that each of us can be a conscious participant in our own evolution. He validates the idea of *conscious creativity,* a concept I have long championed as one of humanity's most powerful tools for change. In *Spiritual Intelligence,* he shows us how we can actively shape the neural networks of our brains, cultivating resilience, empathy, gratitude, and wisdom. This potential is profound—not only can we remake ourselves, but in doing so, we can collectively remake the world.

In *The Possible Human,* I explored how accessing the inner landscape can reveal hidden knowledge and deeper insight. Dawson builds on this by demonstrating that our potential is not limited by age, culture, or background. Through neuroscience-grounded practices, he shows us that we have the creative agency to evolve continuously. His work offers a way to manifest what has been called "cosmic consciousness," aligning our minds with the intelligence fields of the universe itself.

Another significant theme in Dawson's work is the emergence of a new global consciousness. As we evolve, we become increasingly aware of our interconnectedness with each other and with the planet. This awareness is rooted not only in the wisdom of ancient spiritual traditions but in the structures of our brain and mind, as Dawson reveals. These practices allow us to recognize our unity with others and the cosmos, transcending boundaries of culture, nation, and creed.

Dawson's *Spiritual Intelligence* is informed by a reverence for cross-cultural wisdom, integrating timeless truths from diverse traditions to unlock our potential. Drawing on the teachings of master meditators, Indigenous healers, and Western scientists, he reveals that human evolution is a collective journey toward wholeness. As a synthesizer, he weaves together ancient wisdom and modern science to create a tapestry of understanding that speaks to the heart, mind, and spirit.

Moreover, Dawson brings to light how practices like imagination, intuition, mindfulness, and meditation can help us explore our subconscious and access insights that would otherwise remain hidden. These methods are more than tools for personal exploration—they are pathways to a broader awareness with the potential to transform both the individual and the collective psyche. In exploring the "awakened experience," Dawson's book serves as a treasure map for navigating the inner landscape, guiding those who embark on the adventure of self-discovery and self-realization.

Spiritual Intelligence also speaks to the concept of entelechy, the "formative principle" that I explored in *The Possible Human*. This principle, as Dawson explains, is not theoretical; it is a dynamic force within us, urging us toward our fullest expression. His discoveries align with the idea that each of us has a unique potential encoded within our being, and by engaging in practices that awaken our higher capacities, we activate this potential, allowing it to shape our lives and ultimately the world.

In bringing this book to life, Dawson has created a resource for those who feel the stirrings of their potential and sense that there is something more waiting to be awakened. He invites us to step into the fullness of who we are, to explore the vast inner landscape of our minds and spirits, and to become conscious participants in the unfolding cosmos. This path honors the wisdom of the past while carrying us forward into a positive future.

As we face the unprecedented challenges of our time—climate change, social upheaval, and technological acceleration—there is an urgent need for new visionaries, guides, and tools. *Spiritual Intelligence* meets this need, offering us a way to evolve not only for personal fulfillment but for the betterment of the world. This book is more than a guide; it is a call to action, transformation, and evolution.

To those who feel the stirrings of potential and sense an awakening within, Spiritual Intelligence offers a clear path forward. It invites us to step boldly into the next stage of human evolution and realize, together, the beauty and power of the possible human. Let us answer this call with courage, compassion, and creativity—for ourselves, each other, and the world.

In Dawson's articulation of *Spiritual Intelligence,* I see the flowering of ideas that I have explored across decades and cultures. His work brings together insights from spiritual masters and scientific pioneers, creating a cohesive understanding of our shared human potential. His synthesis of ancient practices with modern neuroscience offers a path to unlocking this potential, showing that our evolution is not bound by time or tradition. Rather, it is an ever-unfolding journey toward wholeness, embracing all who wish to join.

Immerse yourself in the wisdom and wonder of this beautiful book. Let Dawson's insights guide you, his practices inspire you, and his vision of the possible human awaken your sense of purpose and possibility. In embracing our Spiritual Intelligence, we are not only transforming ourselves; we are shaping the very future of the planet.

Chapter 1

Your Brain Is Hardwired for Spiritual Intelligence

FORGIVING THE UNFORGIVABLE

Anthony Ray Hinton spent 30 years on death row for crimes he did not commit.

He was accused of two murders and a nonfatal shooting in Birmingham, Alabama in 1985.

The prosecutor had a history of racial bias. He pressed charges despite the fact that Hinton was working in a locked factory 15 miles away at the time and a polygraph test administered by the police had exonerated him.[1]

Much of Hinton's time in jail was spent in solitary confinement. The cell was five feet by seven feet and he was only allowed a single hour of exercise per day.

Yet he became a trusted friend of everyone he had contact with, from other inmates to the death row prison guards.

For more than 15 years, lawyers from the Equal Justice Initiative asked that Hinton's case be reexamined, but Alabama's attorney general declined to do so, ignoring the entreaties of even his guards.

Eventually, Hinton's case reached the US Supreme Court. The justices overturned his conviction unanimously. His first words, as he walked out of the Jefferson County Jail, were, "The sun does shine."

In a later interview, Hinton said: "One does not know the value of freedom until it is taken away. People run out of the rain. I run into the rain…I am so grateful for every drop. Just to feel it on my face."

Later, reporter Scott Pelley interviewed Hinton on the TV show *60 Minutes.* Pelley asked Hinton whether he was angry at the many Alabama officials who had kept him in jail for three decades despite ample evidence of his innocence. Hinton replied that he forgave them all.

Pelley persisted, saying, "But they took 30 years of your life—how can you not be angry?"

Hinton responded: "If I'm angry and unforgiving, they will have taken the rest of my life."

1.1. Anthony Ray Hinton.

In another interview, Hinton is quoted as saying: "The world didn't give you your joy, and the world can't take it away. You can let people come into your life and destroy it, but I refuse to let anyone take my joy. I wake up in the morning and I don't need anyone to make me laugh. I'm going to laugh on my own, because I have been blessed to see another day, and when you're blessed to see another day, that should automatically give you joy."

Looking back at the changes his ordeal had produced, Hinton said, "My faith got stronger."

Your Inheritance from Your Ancestors

What was it about Anthony Ray Hinton that gave him the power of forgiveness? The gift of joy? The ability to rise above his circumstances? Greater faith in response to injustice?

There are people who suffer trivial wounds and slights yet nurture their grievances for decades. They remain consumed by anger the rest of their lives and die holding their bitterness close.

Others, like Anthony Ray Hinton, suffer the most extreme cruelty, yet have the power to liberate themselves from its aftereffects. What is the difference between the two?

Spiritual Intelligence.

Spiritual Intelligence is what connects your individual human consciousness with a consciousness greater than yourself. It makes the difference between living a life rooted in the past, clinging to bygone resentments, and setting yourself free to enjoy the gifts of the present.

What sets these two experiences apart? The answer lies in the way *Homo sapiens* evolved. Our brains, genes, hormones, and neurotransmitters have been shaped by millions of years of experience. Spiritual Intelligence can only be understood in the context of that history.

You are here today, thanks to your ancestors. They did something that resulted in you existing: They survived.

They might have been amazingly wonderful people. They might have been horrifically bad people. They may have had extraordinary gifts, skills, and abilities. They may have been dull and boring. They may have been people of little talent, intellect, or ability—but all of them accomplished one thing.

They survived long enough to produce offspring. They passed their genes along to the next generation. The next generation—good, bad, or indifferent—also did that one essential thing. *They survived.*

Rinse and repeat this process for thousands of generations and here *you* are. You are the result of the survival skills of thousands of earlier humans.

Your ancestors survived partly because of their physical skills and dexterity. Nature rewarded those who were physically fit. Imagination and wisdom helped—"Let's take the tribe back to that valley where we found those delicious sweet potatoes."

But they survived primarily because they were good at one or more of four skills: fight, flight, freeze, or connect.[2]

Connecting means socializing successfully. When danger threatens, you can call on your friends to come to your aid. Sheer brute physical strength is great in a fight or for running away. Freezing—staying so still the enemy doesn't notice you or leaves you for dead—was a last resort, but it could also work to keep you alive.

1.2. An orangutan mother kissing her adult daughter.

Whether it was intellectual cunning, physical strength, or social skill, your ancestors passed these genetic and epigenetic characteristics to the next generation, selecting for those that were most useful for survival. A bigger brain was one of these advantages.

Most species evolve slowly over the course of thousands of generations by natural selection and random genetic mutation. Before early humanoids grew big brains, they also evolved slowly.

Mammals evolved from reptilian ancestors more than 178 million years ago.[3] That's over 100 million years before the extinction of the dinosaurs. It took another 50 million years after the extinction, give or take, for the first primates to evolve. A variety of early hominid species began to appear in the fossil record about 10 million years ago.

The earliest examples of our species, *Homo sapiens,* date from less than 300,000 years ago.[4] But we've taken evolution on a wild ride since then, as the human brain has reshaped the world.

The Explosive Growth of the Hominid Brain

When our brains began to grow, they evolved at an astonishing speed. Check out the chart below. Cranial capacity—the amount of space inside the skull—went from around 500 cubic centimeters three million years ago to 1,350 cubic centimeters today.[5] The greatest increase in size kicked off only around 800,000 years ago.[6]

Most of that brain growth has been in the cortex, the most recently evolved part of the brain. It handles high-level processes such as consciousness, thought, reasoning, language, memory, and emotion.[8] By volume, the cortex makes up 82% of the brain.[9]

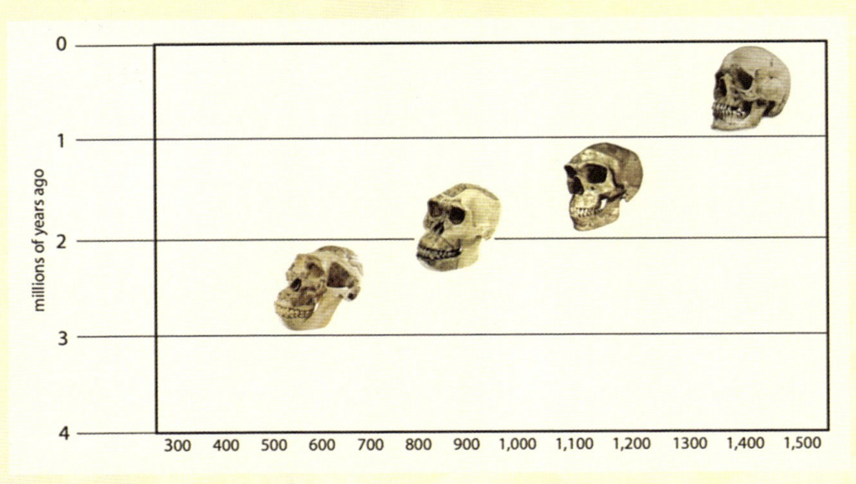

1.3. Brain capacity in cubic centimeters before and after humans began using tools, a span of three million years. Left: Australopithecus africanus. Center left: Homo habilis. Center right: Homo erectus. Right: Homo sapiens.[7]

The Survival Advantage

These recently evolved functions give us *a huge survival edge* when paired with ancient brain structures like the cerebellum, which controls movement, sleep, digestion, respiration, and reproduction.[10]

A great deal of that cranial capacity is devoted to survival. The brain has overlapping networks of regions that have developed over the centuries in response to our survival needs. A large percentage of the volume of the human brain is engaged in the business of keeping us alive and kicking.

I call this neural material inside our heads "Caveman Brain." Caveman Brain is the hardware and associated software that allowed your ancestors to survive. It is superbly good at that skill and it's the reason you are here today.

When Caveman Brain was evolving in that archaic environment thousands of years ago, the world was full of threats. Small bands of early humans roamed the plains hunting and gathering. They were in direct competition for resources with other bands. They were also in competition with other hominid species like Neanderthals and Denisovans.[11]

They were competing for food and water with other species too, includ-

1.4. Reconstruction of the features of an elderly Neanderthal.

ing wolves, tigers, boars, gorillas, and other animals that occupied a similar ecological niche.

Simply surviving from one day to the next was a challenge. Average life expectancy was around 25 years, though if a person survived through the teen years, their chances of longevity improved.[12] But 27% of children died before their first birthday, while 48% failed to survive puberty.[13] The world they lived in was a tough and unforgiving place.

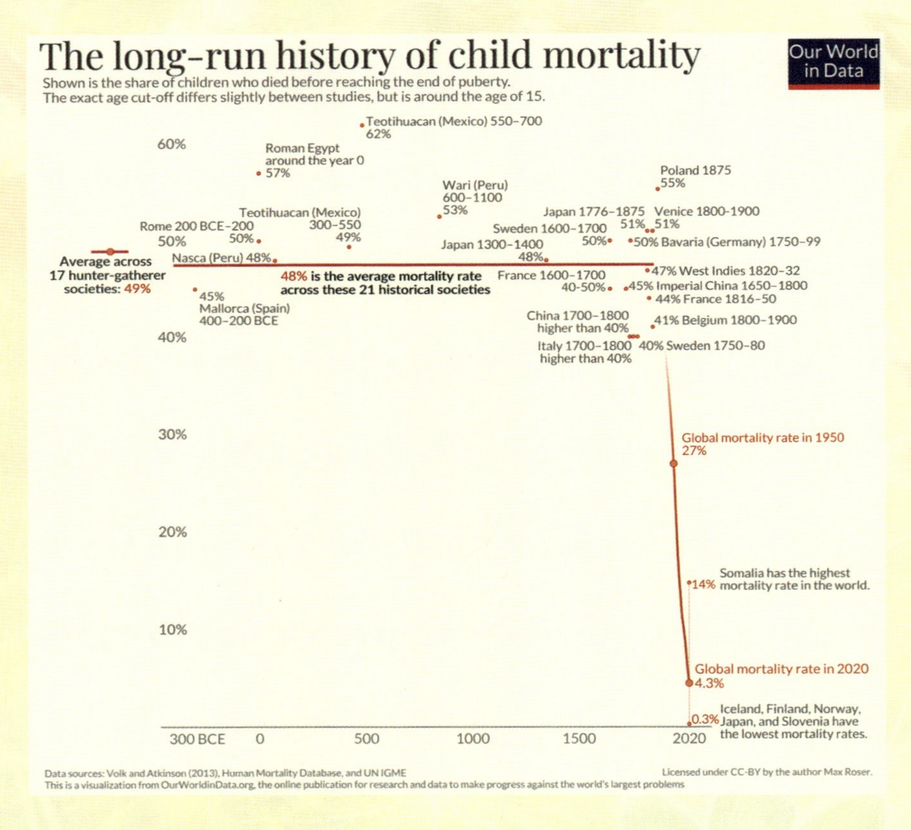

1.5. Mortality rates of children over the last two millennia.[14]

That's why Caveman Brain placed a premium on survival. As humans and their bigger brains evolved, those parts of the brain that contributed to survival commanded a premium. Brains that had an enhanced ability to *detect and evade threats* passed those characteristics on to the next generation.

GUG AND HER CYP17 GENE

In your ancestral line 100,000 years ago were a pair of sisters. Their names were Hug and Gug. They were both young teenagers, born less than a year apart, and they looked so much alike that other members of the tribe could barely tell them apart.

But psychologically, they were polar opposites.

Hug was affable, easygoing, happy, and optimistic. She always saw the bright side of people and events, and her fellow tribe members enjoyed her bright smile, funny comments, and peals of laughter.

Gug, on the other hand, was a born curmudgeon. Paranoid and grumpy, she looked at everyone and everything with suspicion. She was always ready to see the dark cloud inside every silver lining. Gug became angry at the drop of a feathered headdress. She pointed out the faults of everything and everyone. This made her the least popular person in the village.

Little did their Caveman Brains know, but Gug had a rare genetic mutation. It triggered rapid transcription of a gene called CYP17, which makes cortisol, the body's main stress hormone.[15]

The reason for Gug's pessimism was that her cortisol response turned on a nanosecond earlier than it did in other members of her tribe.

Hug and Gug's main job was carrying gourds filled with water from the nearby stream up to the village. They made the trip along the same path through thick jungle several times a day.

Gug saw danger at every turn and constantly imagined threats where there were none. Her paranoid brain pictured tigers lurking in the underbrush, pythons looped over the branches, and black widow spiders dropping from the trees. Hug reassured Gug; after all, they'd done the same trip thousands of times without incident.

But one day there really was a hungry tiger lurking in the bushes. The girls heard a growl as the leaves parted and the tiger charged. Hug and Gug dropped their gourds with a scream and fled in terror.

Because of her genetic mutation, Gug's cortisol response turned on just that nanosecond earlier than Hug's. That gave her the survival edge. She reached the village to sound the alarm. Hug became the tiger's lunch.

Hug's sunny optimism was lost to the human race along with her genetic information. Gug went on to spawn a tribe of little Guglets who all carried her CYP17 gene mutation and shared her miserable disposition. The most paranoid of Gug's children survived. Repeat the process for 10,000 generations and you have a modern human being with a superb stress response.

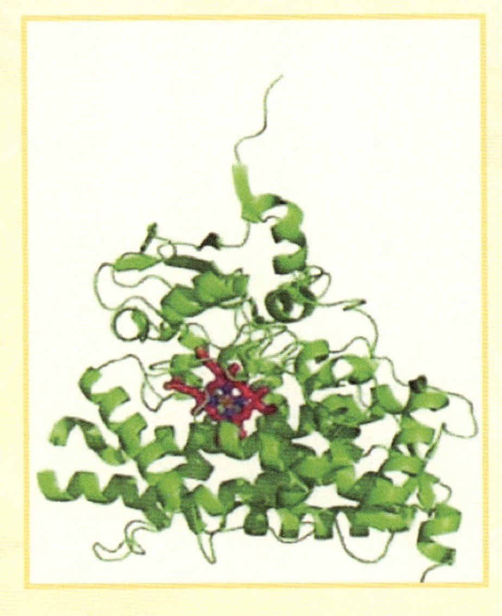

1.6. CYP17 gene.[16]

A brain that *imagines* a threat 10,000 times *in error* is more likely to notice the one threat that's real. A brain that *dismisses* the thousands of imaginary threats is more likely to *miss* the one rare actual threat and get weeded from the gene pool. The long evolutionary process makes the human brain extraordinarily good at identifying the remotest possibility of threat and favors genetic mutations that enhance this ability.

External and Internal Threats

Not only do genes and hormones contribute to survival ability, but clusters of neurons do too. Those parts of Caveman Brain that identified threats and spotted opportunities were favored. These brain regions devoted to survival grew bigger over time.

All this wiring is still inside our heads. That big chunk of neural material that ensured our survival is still there in your brain and mine.

What's changed is our environment. Neanderthals aren't around any longer. As the dominant species, we've displaced the wolves and tigers and we don't have to compete with them anymore. We've driven competing species to extinction and many remaining species are endangered.

We no longer face threats to our existence on a daily basis. Most people in most countries are reasonably safe and secure and don't have to fret about survival needs every day.[17] You don't need to worry about whether you will find water to drink or food to eat today. You don't have anguish about whether you will have a roof over your head tonight and a dry place to sleep. You don't have to worry about a sneak attack from a neighboring tribe or staring down a hungry bear.

What Caveman Brain Does Today

These were the external threats about which your ancestors had to obsess every moment of every day and which drove the evolution of the human brain. That's how you wound up with much of your cranial capacity devoted to survival. But what's missing from your world is all those threats your ancestors faced.

So what does your Caveman Brain do?

It worries. In the absence of *objective* threats to your survival, it obsesses about *subjective* threats. These are things that aren't *actual* threats to your survival but which nonetheless are cause for concern. The economy, job security, unscrupulous corporations, being late, income inequality, the weather, housing prices, inflation, school quality, political candidates, the stock market, retirement benefits, cryptocurrencies, mortgage rates, social media, fake news, real news. The list goes on forever.

Your Caveman Brain can't distinguish between *real* threats and *imaginary* ones.[18] Its stress response is similar whether the threat is an *objective* one like a hungry tiger or a *subjective* one like your next performance review. In the absence of real objective threats, we worry about subjective ones. Worry is in fact our brain's *default* state.

Two brilliant Harvard psychologists, Matthew Killingsworth and Daniel Gilbert, did a study of this phenomenon. They had people download an app onto their smartphones. At random intervals, the app asked participants what they were doing and how happy they were. Eventually, the investigators had over 200,000 data points to work with.[19]

They found that when people were hanging out with friends and family they were usually reasonably happy. Contrary to what you might expect, they were also usually happy at work.

The surprising finding of the study was that when people weren't doing anything, they usually weren't happy. You'd think that, freed of demands and having a break in your day, you'd be enjoying your happiest time. In fact, the opposite turned out to be the case. People were generally least happy during their down time.

The Task Positive Network

When the mind doesn't have anything to do, it *wanders* and typically focuses on unhappy thoughts. Doing things, on the other hand, engages our attention by activating a set of brain regions called the Task Positive Network, or TPN.[20]

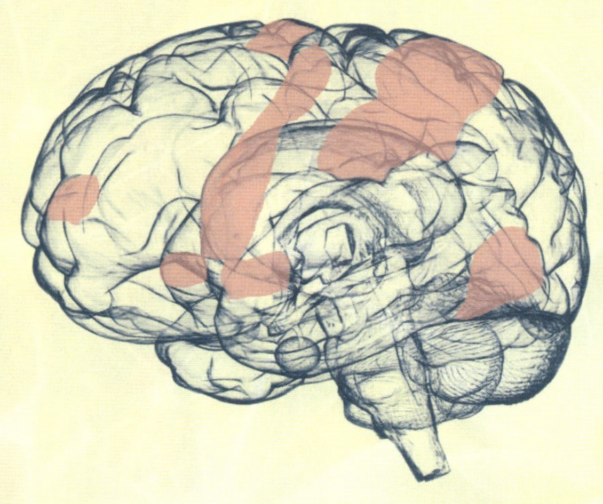

1.7. The Task Positive Network.[21]

When we're solving problems, whether at work, in our family life, or in a video game, the Task Positive Network is engaged and we're generally feeling content.[22] But when the brain doesn't have much to do, the Task Positive Network dials down. A completely different set of brain regions lights up. This set is called the Default Mode Network. It is the state to which our brain *defaults* when we aren't *actively engaged in a task.*

It would be wonderful if our brain defaulted to happiness, but that's not the way evolution designed us. *Our brains default to worry.* When we don't have a task to focus on, our attention wanders and we tend to think about the memories and situations that bother us.

Your Brain's Default Setting

That default setting was incredibly useful to our ancestors. Ruminating about the tiger that almost gobbled you yesterday and reliving the memory in vivid detail revived all the terror of the experience. But it made you much more likely to avoid the tiger tomorrow.

The Default Mode Network is obsessed with the threats that imperiled our survival in the past. It then projects those fears into the future to imagine how we might survive a similar threat tomorrow.

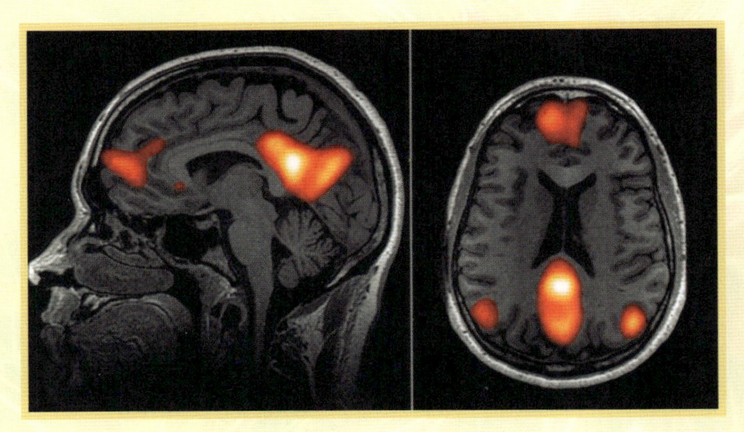

1.8. An MRI image of the Default Mode Network. The front section is the medial prefrontal cortex. The large back section is the posterior cingulate cortex.

On a cold winter morning, when I wake up to meditate at 5 am, I turn up the thermostat. Its nighttime default is 50 degrees Fahrenheit and it defaults to that setting every day from 10 pm to 6 am. I turn it up to 72, but at 6 am its default daytime setting of 70 kicks in and erases my input.

Just as the thermostat in your house returns to its default settings each day, the Default Mode Network is the setting to which your brain defaults.

Our brains use any available spare capacity for rumination and projection when we aren't actively doing a task using the TPN. This system of automatically activating the Default Mode Network worked so well for our ancestors that this hardware grew to make up a large percentage of our brains.

When we try to use our down time to meditate, think positive, or find inner peace, we're swimming against the tide of millions of years of evolution. Your brain is simply not built that way. It's built to *default to worry.* We worry as individuals, we worry as families, we worry as communities, and we worry as a global species. Our media is full of worry, because it's a reflection of our individual worries.

Logically, using your idle time to worry doesn't make sense. It would be much more productive to use it for happiness. But we don't. We deprive ourselves of joy as our minds wander to worry. This isn't because we're morally or intellectually defective. It's simply the way our brains are wired.

We're modern humans living in a world with, for most of us, few threats to our personal survival. But we all have hardwired Caveman Brains. Most people are born, live their lives, and die with brains that default to survival mode in every idle moment. The brain's "negativity bias" has been noted by psychologists for decades.[23]

The Default Mode Network is highly active in people with major depressive disorder, as they default to ruminating on the insults of the past and the stressors of the future.[24] The more the mind is consumed with negative thinking, the greater the activation of the Default Mode Network.

This network isn't a one-dimensional villain, robbing you of happiness. It also performs a variety of useful tasks. It's the part of the brain that builds your sense of self. If you feel safe and secure, enjoyed a trauma-free childhood, and have a healthy sense of self-esteem, contemplation of the past or future may not bring suffering.

The Default Mode Network also handles many routine activities in the background, so you don't have to think deeply about tying your shoelaces or brushing your teeth. Even its mind-wandering setting can produce stream-of-consciousness daydreaming and lead to leaps of creativity.[25]

For most people, however, activation of the Default Mode Network leads to whining, repetitive, self-obsessed misery. Researchers Killingsworth and Gilbert sum it up elegantly when they state, "A human mind is a wandering mind, and a wandering mind is an unhappy mind."[26]

Seeking Happiness

So with the odds stacked against you by millions of years of evolutionary brain anatomy, how do you find happiness?

1.9. Participants at a retreat I taught at Esalen Institute in California.

When we recognize that survival thinking keeps us miserable, we likely aspire to a happier life and take action on that insight. We might take meditation classes, read books on positive thinking, hire life coaches and psychotherapists, go on religious retreats, visit ashrams, listen to inspirational podcasts, or enroll in online programs on positive psychology. For a short while, these do make us happy, but the effect rarely lasts.[27]

Our brains weren't designed to be happy. Our brains are hardwired to default to survival mode. We close our eyes to meditate, but we immediately worry about our To Do lists and email inboxes. We take a walk in the park, but our minds are churning on last week's meetings and next week's deadlines, and we return from our walk having failed to notice the beauty around us. We go on vacation, but mentally we're obsessing over problems at work. We activate our phones intending to read Eckhart Tolle but choose Google News instead.

That's Caveman Brain and that's how it robs us of enjoying our lives. When we take time off, we should be lost in happiness. Instead, our brains default to thinking about the bad stuff to make sure it doesn't happen to us again.

HARRY AND THE FRENCH CONNECTION

Every year, I fly to Europe to teach. My flight starts out at the International terminal at SFO, San Francisco Airport. Passengers have to arrive at the airport a couple of hours early and, after checking in, they wait in the International Lounge. As I sat there one year, I could not help overhearing the conversation of a group of people just behind me.

A man named Harry was traveling with his wife and two friends to Paris, just as I was. Harry was regaling his companions with stories of all the disasters of his previous trip. He told them how the flight was delayed, about how cramped it was, and the neglect he'd suffered at the hands of the flight attendants. The food on the plane had been terrible.

When he disembarked at Charles de Gaulle airport in Paris, he and his companions couldn't find a taxi big enough to accommodate all their luggage. When they did, the driver went to the wrong address and was rude to them along the way. When the driver finally found the correct address, he refused to refund them for the mistaken detour.

Harry got indigestion from the next day's breakfast.

He went on to describe details of the following day and how again everything went wrong. I could hear the emotions in his voice: anger, frustration, contempt, disgust, resentment, blame.

Eventually, I just didn't want to be in that energy anymore, so I left my seat and moved out of earshot. Harry's whole trip to France—as he described it to his companions—had been one miserable disaster after another.

"Why are you going back this year if you had such a miserable time last year?" I wondered.

Surely, Harry must have enjoyed many great experiences—the exquisite cuisine, the wonderful hospitality of the French people, the glorious culture that is France, the stunning beauty of both city and countryside, the ancient monuments, and all the other wonders of the nation.

1.10. France is full of beauty and grace.

Yet as Harry reviewed his trip in his mind, all he remembered was the bad stuff. All the delights of the previous journey had been entirely eclipsed. As his Default Mode Network did its job, Harry remembered his previous trip simply as one bad thing after another.

Think back to Anthony Ray Hinton's story. He'd faced the worst of injustice, neglect, and deliberate cruelty. But with Spiritual Intelligence, he'd made himself happy despite it all.

Harry, on the other hand, faced no life-threatening emergency, yet was able to make himself unhappy over trivial events. That's the difference in quality of life that Spiritual Intelligence confers.

Our brain's negativity bias is a pressing problem for modern human beings. It stops us from being happy. And the challenge we all face is how to have a happy life when many of the neurons inside our skulls are determined to keep us in misery.

Positive thinking is no match for Caveman Brain. Meditation rarely brings relief, as any aspiring meditator will tell you.[28] Vowing we won't behave in the same stupid way tomorrow as we did today doesn't usually work. New Year's resolutions to be better, kinder, more compassionate, wiser, and more positive rarely last past January 15.

How to stop suffering and be happy is the central dilemma that philosophers, scholars, priests, doctors, and spiritual teachers have been trying to solve for millennia. Anxiety and depression statistics show that we're losing ground.[29]

Neuroscience Changes Everything

I've been wrestling with these big issues for many years. When I was a teenager, in an effort to escape the misery of Caveman Brain, I went to live in a spiritual community. I studied the common teachings of the world's great religions, what philosopher Aldous Huxley called "the Perennial Philosophy."[30] We grew organic food, practiced energy techniques like Reiki, and meditated sporadically.

None of these practices made me much happier. I studied with the great human potential teachers of the 1970s and eventually graduated from Baylor University still only marginally more content.

But in my forties, I made a pledge to meditate daily. Within a few months, every part of my life changed. Money, family, career, and health all began to shift. I discovered Energy Psychology and rapidly released the weight of my past. With colleagues at various universities, I began to research these methods. I founded the National Institute for Integrative Healthcare and, to date, we've led or contributed to over 100 studies published in peer-reviewed medical and psychology journals.

Just after the turn of the century, I worked with Bruce Lipton on publishing his groundbreaking book *The Biology of Belief*.[31] At that time, science was captivated by the new science of epigenetics. Studies had shown that diet and other environmental factors could turn genes on and off. This led me to write *The Genie in Your Genes*.[32]

The Genie in Your Genes

In the book, I predicted that spiritual, emotional, and psychological experiences were likewise epigenetic. Hundreds of studies eventually confirmed this effect. Research performed by my colleagues and I showed significant changes in immunity and inflammation genes when consciousness and energy changed.[33] I developed a simple method called EcoMeditation that uses seven evidence-based practices to bring people into deep states fast.

Prior to this, psychologists, as well as researchers like me, largely assumed that the best we could do was to make peace with Caveman Brain. We needed to resign ourselves to the fact that our brains were built for survival, so our recourse was to deactivate the fight-or-flight response using stress-reduction methods.

The science prevailing at that time regarded Caveman Brain as an inescapable fact. The best option we had was to modify our behavior. If you have a car with poor wheel alignment and the steering pulls to the right, you have to correct for this defect every moment of every drive. You drag the steering wheel slightly to the left.

Same approach to Caveman Brain. Psychological research told us that the most we could hope for was to correct for the brain with which we'd been born.[34]

A decade after *The Genie in Your Genes,* I wrote *Mind to Matter: The Astonishing Science of How Your Brain Creates Material Reality.*[35] In it, I reviewed over 400 studies. They show that as your consciousness shifts, different molecules are produced inside your body and brain. Other research shows that you affect the physical world around you.

By that time, there was a lot more research available on the brain, especially the brains of advanced meditators and spiritual adepts. I began to practice the methods they used, layering them into EcoMeditation. These led me to extraordinary states and I found myself becoming more happy than I had ever imagined possible.

1.11. A morning meditation shortly after the fire.

I wrote my next book, *Bliss Brain,* after a series of disasters in my life.[36] Chapter one tells the story of how my home in California was destroyed by a huge wildfire. My wife and I escaped moments before the flames consumed everything we owned. That night, 5,400 homes were destroyed and 22 people lost their lives.

The subsequent months were chaos. A series of financial and health disasters followed the fire. I lost my entire retirement savings as I struggled to keep my business afloat. I had to undergo surgery after a debilitating physical injury.

Yet every morning, during meditation, I found myself in states of such extraordinary bliss that they defied comprehension.

I began to do research on the brains and minds of people exercising their Spiritual Intelligence.[37-39] I read dozens of studies of monks, nuns, and adepts having "enlightenment" experiences. I networked with other experts in the field. The collective evidence revealed that the brain regions active in these people and the brain waves they produce are those of Bliss Brain.

The Enlightenment Network

Not only do these adepts experience extraordinary happiness, their brain function is different. And it's different in ways that are consistent day after day and person after person. As Caveman Brain dials down—as people move out of fight or flight—and into bliss, a completely different set of brain regions lights up.

We can group these regions into *four distinct circuits:*

1. The Emotion Regulation Circuit
2. The Attention Circuit
3. The Selfing Control Circuit
4. The Empathy Circuit

Together these four regions make up the Enlightenment Network. When advanced practitioners are meditating, these four circuits light up in their brains, working together to produce a blissful experience.

In 1995, Harvard psychologist Daniel Goleman redefined the scope of human consciousness in his groundbreaking book *Emotional Intelligence*.[40] Goleman is a lifelong friend of neuroscientist Richard Davidson, who pioneered the use of MRIs to measure the brain states of meditators. Together they wrote the book *Altered Traits: Science Reveals*

How Meditation Changes Your Mind, Brain, and Body.[41] In it, they review the many studies of adepts and identify these four circuits.

The monks whose brain activity they study have spent more than 10,000 hours in meditation over the course of their lifetimes. Some have meditated for over 50,000 hours.

Andrew Newberg is a neuroscientist and coauthor of *How Enlightenment Changes Your Brain: The New Science of Transformation.*[42] One extensive set of studies Newberg conducted examined the brain activity of Franciscan nuns who've spent decades in contemplation and spiritual practice. It identifies a pattern of brain activation similar to that of the Tibetan Buddhist monks in Davidson's studies. This pattern is common to masters of every spiritual tradition, from Qigong to Sufism to Kabbalism to shamanism.

Training Bliss Brain

When we put these modern-day saints into MRI scanners, we find low activity in the Default Mode Network and high activity in the Enlightenment Network. The Enlightenment Network suppresses the activity of the Default Mode Network.[43]

1.12. With advanced scanners like this Siemens fMRI, scientists can identify which brain regions are active and how they communicate with each other.

Adepts like the Franciscan nuns and Tibetan monks have developed this skill over the course of many years. It shows that *with practice we can learn to regulate Caveman Brain by activating the Enlightenment Network.* One famous example is the Dalai Lama's former translator, Matthieu Ricard. MRI scans show that the Bliss Brain regions of his brain are so active that a *Smithsonian Magazine* article called him "The World's Happiest Man."[44]

The implications of these neuroscience discoveries are stunning. They are a game-changer in terms of both individual human happiness and the wellbeing of the human species. We no longer need to live trapped in fight or flight by Caveman Brain. We can learn to turn on the Enlightenment Network and live generally happy lives.

Best of all, MRI research has shown that you don't have to be a renunciate monk or nun to achieve Bliss Brain. With psychologist Peta Stapleton and neuroscientist Oliver Baumann from Bond University, I performed a randomized controlled trial with novice meditators. We found that after using EcoMeditation for just 22 minutes a day for 28 days, their brains showed similar patterns of activation.[45]

As you'll learn in Chapter 5, anyone can be trained to reach Bliss Brain quickly. My colleague Judith Pennington found electroencephalograph (EEG) changes after the very first experience of EcoMeditation.[46]

The resilience that took me through the fire and allowed me to enter Bliss Brain every day even while disasters were unfolding around me is available to every human being. My passion for sharing this discovery motivated me to write *Bliss Brain* and make EcoMeditation freely available to anyone who wanted it.

Millions of people have now been learning to get into Bliss Brain. We're researching their experiences in order to make these skills universally and freely accessible, as you'll see in Chapter 5.

The skills of Bliss Brain are as trainable as any other skill. Just as you can teach yourself to speak Swahili, play chess, or ride a bicycle, you can train your brain to make you happy. You can practice turning on the Enlightenment Network until it becomes a habit.

Your Brain Anatomy and Spiritual Mastery

It is well worth understanding the anatomical structure and the physiological function of each of the four circuits. Millions of people have now learned EcoMeditation and many of them describe noticeable changes in the way their brains feel.

When worries about past and future drop away, for instance, and they come fully into the present moment, many meditators describe a "tingling feeling" in their medial prefrontal cortex. Knowing where in your brain this tissue is located, and the functions it performs, helps you understand the experience.

This is the same area referred to in mystical traditions as the "third eye." Paramahansa Yogananda, author of *Autobiography of a Yogi,* told his disciples to pay particular attention to sensations in this area.[47] Sant Kirpal Singh, a great spiritual teacher of the early 20th century, called it "the bridge between the body and the soul."[48]

Physical sensations in this "third eye" area were also noted by neurofeedback pioneer Les Fehmi, late director of the Princeton Biofeedback Center. Fehmi found that when his students focused their attention on this part of the brain, their alpha brain waves—a characteristic of altered states—expanded automatically and exponentially.[49]

When you understand that the medial prefrontal cortex is part of the brain anatomy involved in the Enlightenment Network, sensations in your "third eye" area during EcoMeditation and other Spiritual Intelligence practices make sense.

The Anatomy of the Four Key Circuits

Below are images of the locations of the four circuits. The only abbreviation I use is PFC for prefrontal cortex; otherwise, the full name of each brain region is used.

The Emotion Regulation Circuit

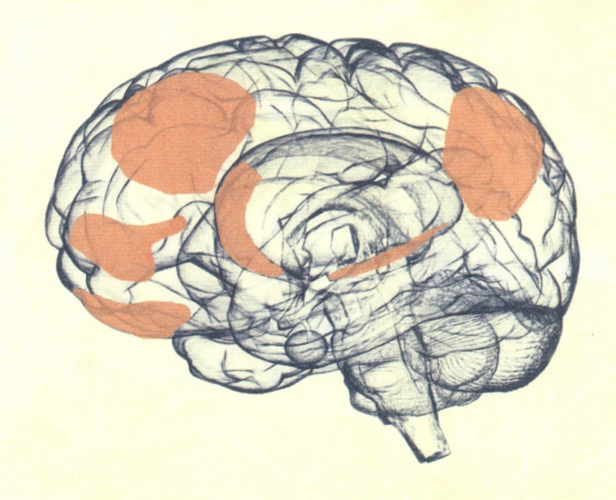

1.13. ***Emotion Regulation Circuit:*** *Voluntary regulation by dorsolateral PFC and ventrolateral PFC; automatic regulation by medial PFC, anterior cingulate cortex, posterior parietal, and orbitofrontal cortex; coordination by dentate gyrus.*[50-51]

In my book *Bliss Brain,* I recommend activating the Emotion Regulation Circuit first. That's because if your attention is captured by the endless succession of emotions that flit across your awareness, it's almost impossible to find inner peace.

If you're reacting to every random feeling of judgment, resentment, criticism, disappointment, anger, disapproval, shame, stress, disgust, and guilt that pass through each day's experience, you don't have the mind-space for the joy and peace you deserve. The Emotion Regulation Circuit regulates all those negative emotions, opening up the space for a greater experience.

The dorsolateral PFC is particularly important. That's the core of the brain's executive center. When you use it to *decide* to regulate negative emotion, it passes those signals to the emotional brain, aka the limbic system, through the ventral PFC. The ventral PFC, especially the ventrolateral PFC and the ventromedial PFC, are well-developed in meditators.[52]

With practice, this emotional control becomes automatic. You aren't just regulating negative emotion during your meditation periods, you're doing it throughout your life. The habit of negative emotion drops away, to be replaced by the habit of equanimity.

The Attention Circuit

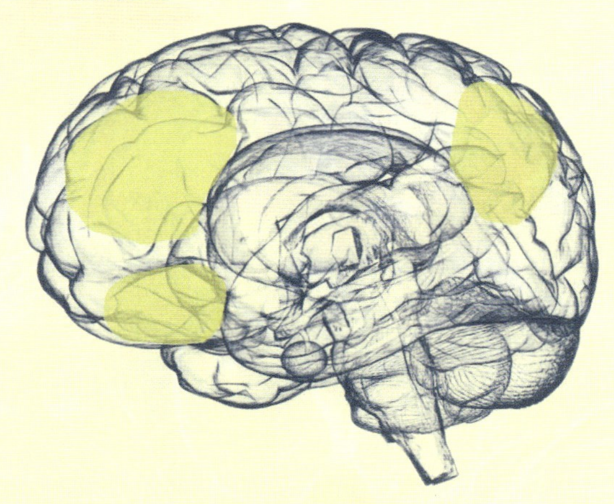

*1.14. **Attention Circuit:** dorsolateral PFC, parietal cortex, ventrolateral PFC, and, deeper in the brain, caudate nucleus and dorsal anterior cingulate cortex.*[53]

This is the circuit that governs the direction of your attention. If your attention is wandering aimlessly or driven by the negative emotions that pass through awareness, you can't focus on what's truly important in life. If instead you activate the Attention Circuit and fixate on the positive, you use the gift of awareness to fill your mind with serenity.

The Selfing Control Circuit

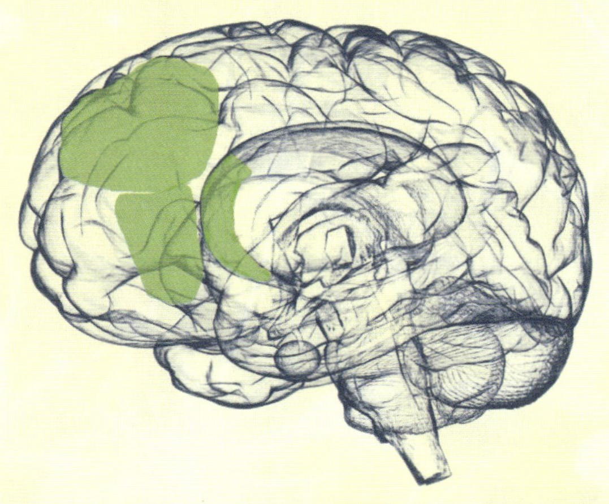

*1.15. **Selfing Control Circuit:** dorsolateral PFC, anterior cingulate cortex, and inferior frontal gyrus.*[54]

Our minds tend to default to self-absorption. That's the endless mental chatter about I, me, and mine. It is the Default Mode Network, especially the medial PFC, that constructs our sense of self.[55] It's preoccupied by what hurt *me* in the past and what might hurt *me* in the future. It's not in the present moment, and that's the place we find the greatest joy.

This endless stream of self-absorbed chatter keeps us out of the present and out of happiness. Researcher Jeffery Martin, whose work we'll study in detail in Chapter 3, calls this "the Narrative Self."[56]

That's because it produces an ongoing commentary about the trivia of what's going on around us and how that affects *me*. In one conversation with me, Martin quipped: "If the Narrative Self were a person, you sure wouldn't want to be trapped in an elevator with him!"

Seasoned meditators are able to regulate this fascination with the Narrative Self. Spiritual Intelligence allows us to drop our obsession with the individual self, separate from the rest of creation. We stop selfing. That allows us to become one with something greater than ourselves.

The part of the brain that spiritually intelligent people use to do this is the Selfing Control Circuit. Quieting your inner commentator catapults you into the present moment and the bliss we find there.

The Empathy Circuit

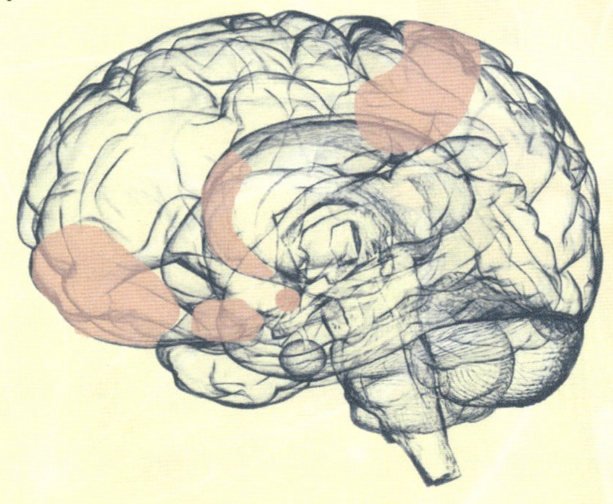

*1.16. **Empathy Circuit:** anterior cingulate cortex, ventromedial PFC, hypothalamus, and somatosensory cortex.[57] Not shown: insula.*

The Empathy Circuit is nature's unique gift to you as a social animal. Its center, the insula, contains special neurons called von Economo neurons. They are found only in highly social species such as elephants, dolphins, and monkeys.[58-59] Your insula enables you to relate to the feelings of others as though they were your own.

Of all the types of meditation that influence the brain, we find that those based on compassion produce the fastest and strongest neural results.[60]

Spiritually intelligent people engage all four circuits. They regulate their negative emotions. They focus their attention. They dial down self-absorbed mind chatter with the Selfing Control Circuit. They engage compassion. These four circuits dance together as the Enlightenment Network.

This hardware is present in your brain. It exists in the brain of every human being. It's *part of your anatomy,* just like your toes, your liver and your biceps. It's as real as the heart that keeps your blood flowing and the lungs that gather oxygen for your cells. Spiritual Intelligence is what allows you to *boot up this awesome neural network you already have inside your skull.*

Shifting Between Networks

Research into these circuits is ongoing. The brain is unimaginably complex and our understanding of it increases continuously. Scientists haven't mapped these brain circuits definitively and our current understanding will change as we understand them better.

Also, if you decide to geek out and read the original MRI studies I reference, you'll notice that different scientists define the regions and circuits differently. Some regions, such as the anterior cingulate cortex, are part of more than one circuit. To complicate the picture further, the connectivity between brain regions—how they talk to each other—plays just as important a role as the structures themselves.

However, current science does allow us to define the broad outline of these four circuits in our brains as well as the ways they communicate with one another, activate, change, and grow.[61]

When you're shifting brain activity into the Enlightenment Network, you are making a conscious choice. That involves mental focus—a task—so your brain does not default to the misery setting of the Default Mode Network.[62] The Task Positive Network *stays lit up enough* to suppress the activity of the Default Mode Network. Even simple meditative tasks like counting your breaths or reciting a mantra can be enough to produce this effect.

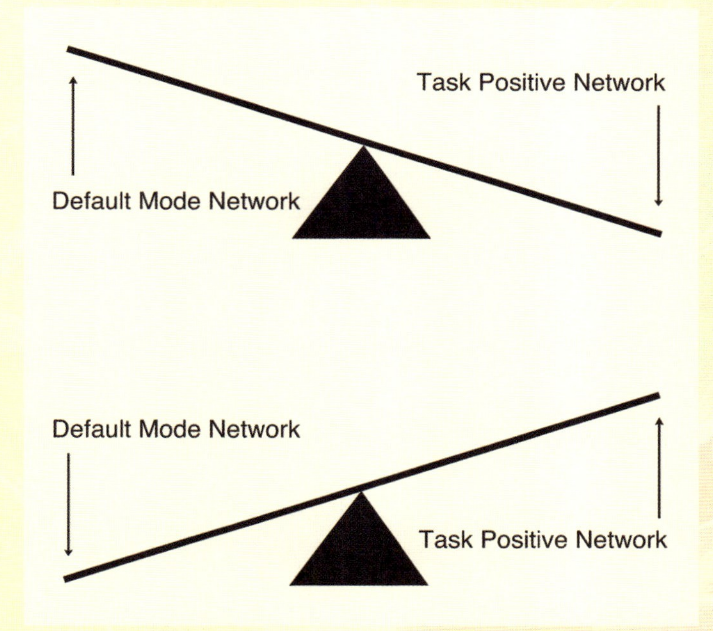

1.17. These two networks operate like a seesaw or teeter-totter. When activity in one increases, activity in the other decreases and vice versa.

Shifting the balance from Caveman Brain to Bliss Brain requires a unique form of intelligence. Intellectual Intelligence or IQ can't do this any more than you can talk yourself into thinking positively. While Intellectual Intelligence is incredibly useful for solving problems, it can't solve the problem of the hardwiring of Caveman Brain.[63]

Emotional Intelligence, which is usually abbreviated as EQ (emotional quotient), is an equally valuable part of being a functional adult human being, as we'll discover in Chapter 2. All the other intelligences—such as somatic intelligence, linguistic intelligence, interpersonal intelligence, mathematical intelligence, and musical intelligence—also play their role.[64]

But it is Spiritual Intelligence or SQ that gives us the insight and motivation to activate the Enlightenment Network.

How Spiritual Intelligence Transforms Subjective Experience

Imagine if Harry had possessed the skill of SQ. The first thing he'd have done was engage his *Emotion Regulation Circuit*, regulating his negative feelings. He'd then have been able to activate his *Attention Circuit* and direct his attention to the great experiences of the previous trip to France.

Rather than a tirade of whiny self-absorption, with the world revolving around Harry—typical of the Default Mode Network—he'd have seen things from the point of view of other people, as the *Selfing Control Circuit* turned on.

That might have led to compassion for others, as the *Empathy Circuit* kicked in. Perhaps Harry would have asked the taxi driver compassionate questions. Harry would then have discovered that the driver was a recent immigrant, unfamiliar with the city, with a family to feed, his rent due that day, and terrified of losing his job.

All the same external events might still have happened, but Harry would have been left with great stories to tell his friends, as well as amusing anecdotes about the misadventures that are an inevitable part of any trip. Same outer experience, but a completely different inner experience.

That's one of the many gifts we get from SQ.

How might the brain regions active in SQ have contributed to the experience of Anthony Ray Hinton?

His *Emotion Regulation Circuit* stopped him getting angry at the prosecutor who put him behind bars, the Alabama attorney general who kept him there for an extra 15 years, and the prison guards.

While his own situation was dire, Hinton's *Selfing Control Circuit* was active enough to suppress his situation from becoming the only problem he focused on.

He used his *Attention Circuit* to direct his attention toward positive influences and other people. His *Empathy Circuit* then kicked in. He mentored other prisoners and formed supportive friendships, even with the men who were assigned to guard him.

Objectively, Harry had a much better life than Anthony Ray Hinton. Yet subjectively, Harry's inner experience was miserable, while Hinton was able to uplift not just himself, but everyone around him. In an interview he said: "When every court was saying 'no,' I believe God was still saying yes. I had to somehow find that faith and reach deep down in my soul and believe in the teaching that my mother taught me as a young boy, that God can do everything but fail."[65]

Spiritual Intelligence Defined

My definition of SQ is a simple and commonsense one: "Spiritual Intelligence is the ability of human consciousness to interact with universal consciousness."

That's what those Tibetan monks and Franciscan nuns we study with MRI scanners are doing. They're leaving behind the earth-bound survival focus of the Default Mode Network, turning on the Enlightenment Network, and entering blissful communion with a consciousness greater than themselves. Even when confronted with the worst of racial prejudice and social injustice, Anthony Ray Hinton's faith was strengthened as he connected with a greater consciousness. SQ is what enables this type of experience.

There is a vast field of academic deliberations about the nature of consciousness and an even vaster one about intelligence.[66-67] Little of it has any relevance to our everyday lives. So in this book, rather than entering this academic debate, I use the commonsense definition of SQ above. I'll also defer questions like whether plants and animals have some form of SQ, whether intelligent life forms elsewhere in the cosmos might possess SQ, and other hypothetical questions, in order to focus on the practical application of SQ in real life.

I also use the word "consciousness" rather than "mind" as other scholars defining SQ do, since the experience of mystics can encompass both passionate emotions and vivid physical sensations.[68-69] It's much more than a purely mental experience. It's also more than a connection, as we'll discover in Chapter 6. It's an active *two-way interaction.*

People Embodying Spiritual Intelligence

Let's get even more concrete. When you think of *a spiritually intelligent person,* who pops into your mind? Mother Teresa? Rumi? Amma, the "Hugging Saint"? Gandhi? Thich Nat Hanh? Martin Luther King? Francis of Assisi? The Dalai Lama? Maya Angelou? Thomas Merton? Wangari Maathai? Gangaji? Chief Joseph? Joan of Arc? Ram Dass? Saint Claire of Assisi? Paramahansa Yogananda? Wayne Dyer? Greta Thunberg? Billy Graham? Nelson Mandela? Hildegard of Bingen? Pope Francis?

1.18. To people of many ages, places, and religions, Mother Teresa has represented an example of SQ.

Seane Corn? Eckhart Tolle? Chief Seattle? Pema Chodron? Moses? Teresa of Avila? Brother Roger of Taize? Khalil Gibran? Kabir? Desmond Tutu? Marianne Williamson? Meister Eckhart? Ramana Maharshi? Brother Lawrence? Paulo Coelho? Ravi Shankar? Don Miguel Ruiz?

Please write down the names of a few people you believe embody SQ. They may be famous public figures or saints, ancient or contemporary. They could be ordinary people you've known personally. Throughout this book I'll be dialoging with you using journaling, exercises, and resources. Recording the names of people who represent SQ is a good way to start.

All of these people drew their inspiration from a higher source. Their "human consciousness," as the definition puts it, has the ability "to interact with universal consciousness." While they came from different continents, epochs in history, cultures and religions, this ability marks them as spiritually intelligent people.

When Anthony Ray Hinton decided to forgive rather than remain embittered against those who had imprisoned him wrongfully, he used SQ. When the people on your list acted and spoke in ways that inspired

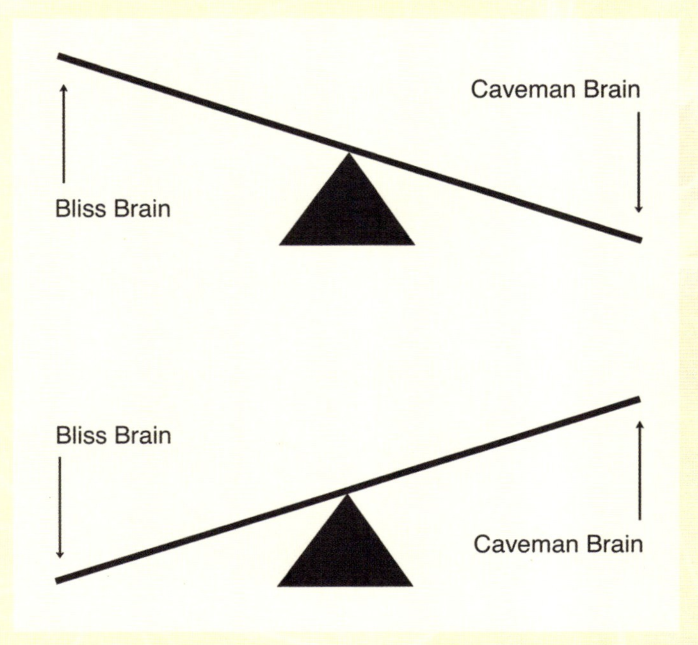

1.19. As we use SQ to enter Bliss Brain, we dial down the stress-focused circuits of Caveman Brain and vice versa.

you, they exhibited SQ. And when you activate the regions of your brain that make you happy, compassionate, grateful, and resilient, you grow your own SQ.

Every Brain Has This Hardware

Neuroscience has changed the way we view SQ, because we now understand that mystical experiences of oneness and bliss aren't just mysterious subjective experiences. They are accompanied by objective changes in brain function, gene expression, and neurochemistry.

We also now understand that we aren't at the mercy of Caveman Brain. Yes, we have that ancient stress response hardware embedded in our brains. But we also have the circuits of the Enlightenment Network. Every single human being has the hardware inside their head to run the software of bliss. Use SQ to turn it on and you create a whole new life for yourself.

The Journey Ahead

I can't tell you how excited I am to share these new scientific discoveries with you in this book. They have radical implications for the direction of your life. What we learned here in Chapter 1 is just the beginning. In the coming chapters, we'll learn about the characteristics of SQ. We'll discover together that it is not only possible to activate the circuits of the Enlightenment Network, but also to make them bigger—and better at sending neural impulses through your brain.

In Chapter 2, you'll be amazed to learn *how quickly you can grow these circuits*. Initially, turning them on takes great focus, but with practice you do it automatically. EcoMeditation makes it easy and I'll give you several free audio tracks to speed your journey.

This takes you to what Chapter 3 calls the "tipping point," where you cross an invisible threshold into lasting happiness. Studies of thousands of people show that this skill is learnable and trainable. Exercises and resources at the end of each chapter build your ability step by step.

In Chapter 4, we'll learn the five characteristics of people who've crossed the tipping point. Based on Andrew Newberg's work in *How Enlightenment Changes Your Brain,* I developed an assessment called the Transcendent Experiences Scale to measure this shift. My research shows that almost anyone can get there given time and effective training.

We'll also delve deep into the experience of the mystics. But we'll discover in Chapter 5 that it doesn't take religion or spirituality to get you there: the "flow" states experienced by peak performers provide *a back door* to SQ. We'll see how SQ transforms our bodies, triggering the expression of health and longevity genes and synthesizing reward neurotransmitters like dopamine and pleasure hormones like oxytocin.

Every chapter tells stories of people who embodied SQ. Some, like Joan of Arc, Jimmy Carter, Albert Einstein, and Pope John Paul II, are household names. Others are little-known but give us shining examples to follow.

In the final chapters, we'll see how the brain change sparked by SQ is *changing society and the world,* and the astonishing contribution SQ has made to historical events over the past 500 years. We'll learn how SQ is leading humankind into an entirely new evolutionary direction, with profoundly hopeful consequences for our species and the planet.

Cultivating SQ has the potential to change your life completely and I look forward to being with you on the exciting journey that lies ahead of us.

Deepening Practices

The Deepening Practices for this chapter include:

- **Journaling Exercise 1:** Who, to you, represents SQ? Make a list in your journal.
- **Journaling Exercise 2:** Has knowing that you have the brain hardware to experience Bliss Brain made a difference in how you see yourself? Record how in your journal.
- **Dawson Guided EcoMeditation:** Activating Bliss Brain

Download a free EcoMeditation track at SpiritualIntelligence.info/1.

Extended Play Resources

The Extended Play version of this chapter includes:

- The seven steps of EcoMeditation
- Calendar of upcoming workshops and live events
- *Slate* interview with Anthony Ray Hinton
- Podcast: The Five Core Characteristics of Enlightenment Experiences with Andrew Newberg and Dawson Church
- Video Interview: Why Meditation Matters: Daniel Goleman and Richard Davidson with Richard Gere

Get the extended play resources at: SpiritualIntelligence.info/1.

Updates to This Chapter

As new studies are published, this chapter is regularly updated. Get the most recent version at: SpiritualIntelligence.info/1.

Chapter 2

Shrinking Suffering and Growing Happiness

Growing Those Muscles

Hit the gym, start pumping iron, keep it up regularly, and your muscles will grow. Bodybuilders, with their giant muscles, are living proof. They are the literal embodiments of manifestation. Some have used their sculpted physiques to launch new careers. One whose meteoric rise to fame has become legendary is Dwayne Johnson. He began as a professional wrestler, and adopted the nickname "The Rock," in 1997.

DWAYNE "THE ROCK" JOHNSON: MORE THAN MUSCLE

2.1. Dwane Johnson.

In February 2022, Forbes magazine cited Dwayne Johnson, then 49, as the highest-paid actor in the world. The former world wrestling star turned action movie icon has a net worth of $800 million from film, television, and his multiple businesses, which include Seven Bucks Productions. According to the Seven Bucks website (sevenbucks.com), the company develops "content rooted in authenticity, strong storytelling, and passion" for movies and television, among other platforms.

The star of the *Fast & Furious* franchise, *San Andreas,* the *Jumanji* films, *Jungle Cruise,* and, in 2022, *Black Adam* (for which Johnson was paid $22.5 million) traces the beginning of his rise to a particular day in 1995. He had just been cut from the Canadian Football League, which he had thought was going to be his path to the American National Football League and a career as a professional football player.

Of that day, Johnson says, "In 1995 I had $7 bucks in my pocket and knew two things: I'm broke as hell and one day I won't be." He turned to professional wrestling, reinventing himself as "The Rock." Just three years later, he won his first World Wrestling Federation (WWF) championship. Later he would name his production company Seven Bucks to remind him to call upon his inner strength during difficult times.[1]

Johnson notes, "I'm always asked, 'What's the secret to success?' But there are no secrets. Be humble. Be hungry. And always be the hardest worker in the room."

Another mantra of Johnson's life has been "No matter who you are, being kind is the easiest thing to do."

His kindness extends to donating millions to charitable organizations, including the Make-A-Wish Foundation, I Have a Dream Foundation, Parkinson Society Maritime Region, Red Cross, Starlight Children's Foundation, and Until There's A Cure. Further, in 2006, he founded the Dwayne Johnson Rock Foundation, dedicated to helping at-risk and terminally ill children.[2]

In both 2016 and 2019, he made *Time* magazine's list of the world's 100 most influential people.

As evidence that his field of influence derives from far more than his impressively muscular body, he urges, "Be kind, be the change, be the difference. Do it! Think it! Be it!"[3]

While Dwayne Johnson has many admirable qualities, his career has been based on his physique. Johnson's kindness is invisible, but his muscles are iconic. That's why bodybuilders go to the gym and work out: to be seen. As you challenge your muscles, they get bigger, and bodybuilding contests are won by the people with the largest muscles. An increase in muscle mass is the obvious visible result of the huge number of hours spent in the gym by these athletes.

Brain Growth Is Real Though Invisible

Our brains can grow as certainly as our muscles can. While the process is invisible to us, brain growth is as real as muscle growth.

Our brains grow at a particularly furious pace from conception to birth. Newborns have over 100 billion neurons. To get there, throughout pregnancy they're adding about 250,000 nerve cells a minute.[4]

While this process is most pronounced during childhood, it doesn't stop once you're grown up. The adult human brain also adds neurons daily. As you go about your normal life, experiencing events, doing your routines, and learning new information, your brain adds at least 1,000 new neurons a day.[5]

Not only do new neurons form, but existing ones create additional connections to communicate with their neighbors. In a study of the brains of rats, after new neurons had been formed, they kept adding connections for six months. That's a quarter of the rat's lifespan. In a human brain, the equivalent time scale is over 20 years. So your brain is both adding new neurons, and growing new connections between old ones.[6]

The rat study also found that the rate of growth of adult neural connections was actually faster than that found in infants. Compare the image on the left with the one on the right in Illustration 2.2. These are images of actual neurons from an adult rat brain over the course of 24 weeks. As you can see, they continue to form new connections. The same thing is happening in your brain every moment, and perhaps more rapidly than when you were a child.

When you look at before and after photos of bodybuilders, you can see the change that exercise produces to their muscles. "I was a 97-lb weakling" proclaimed bodybuilder Charles Atlas in a legendary 1920s advertising campaign. But using his "dynamic tension" method of working out his muscles, Atlas eventually topped 200 pounds. In 1922, after a contest held in Madison Square Garden, the magazine *Physical Culture* declared Atlas to be "America's Most Perfectly Developed Man."

2.2. Growth of new connections in neurons created during adulthood. Left, 2 weeks, center, 4 and 7 weeks, right, 24 weeks.[6]

You can't see the brain growing like the muscles of Dwayne Johnson and Charles Atlas, but it's happening just the same. As you work out specific groups of nerve fibers, they grow new synaptic connections and increase their mass.

Just because you can't see something doesn't mean it's not happening. There are all kinds of processes occurring in your body that you can't see. You can't see your red blood cells carrying oxygen to all your other cells. You can't see your immune system attacking viruses and bacteria that invade your body. You can't see your digestive system extracting nutrients from food. Nonetheless, all these things are happening daily in your body.

But how can the brain be growing daily within the fixed confines of a skull? This capsule of hard bone doesn't get bigger as we age. So how can the brain inside it be getting bigger?

Nature has developed an ingenious method to create a brain that grows within a skull that does not. In adulthood, the brain grows by *folding in on itself.* Think of walking 10 miles or kilometers on a flat surface. Head for your goal in a straight line, and after walking the distance you're at your destination.

Now compare that to walking to a point 10 miles away as the crow flies, but on a surface folded into many hills and valleys. It could take you 20 miles of hard effort to get to that same point that's just 10 direct miles away.

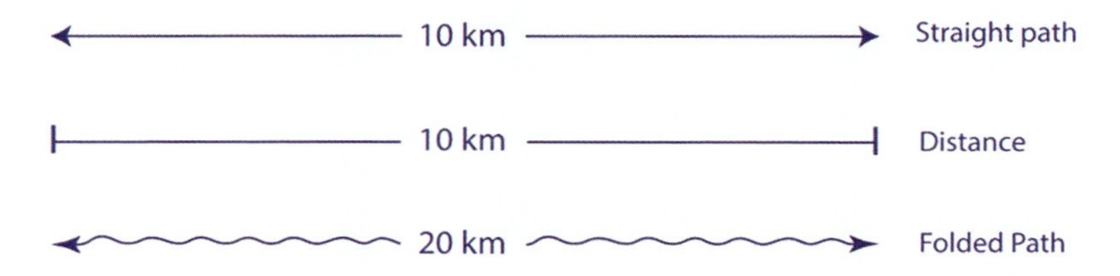

2.3. Folding allows much more information to be stored in a limited area.

The process of human brain folding is called gyrification, and the more adult brain cells you produce, the greater the gyrification of your brain inside your skull.[7] We also have fluid-filled cavities called ventricles in the center of our brains. These also provide space for the expansion and contraction of neural pathways deep inside the brain. This explains how we're able to grow bigger brains inside the fixed confines of a human skull.

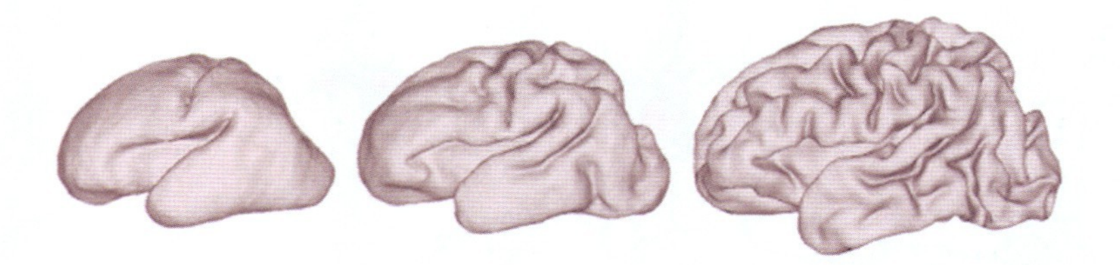

2.4. The brain grows within the fixed limits of the skull by producing additional folds. Left is 27 weeks after conception, center is 31 weeks, right is 37 weeks in a developing human brain.[8]

So while you can't see your brain circuits the way you can see your muscles, the ones you're working out are growing all the same.[9] As you exercise specific parts of your brain, they can grow either slowly or rapidly. Research shows that people with greater cognitive ability and intellectual intelligence show increased gyrification in the executive centers of the brain.[10]

The process continues throughout adulthood. Gyrification means that the amount of new neural material your brain creates can be substantial over the course of your lifetime. As we'll see in Chapter 3, this process can produce dramatic changes in your level of happiness, and in Chapter 4, we'll learn how neural development is linked to four particular characteristics of Spiritual Intelligence.

Brain Shrinkage Is Inevitable

We've all heard the saying, "Use it or lose it." If we use our muscles, they stay strong and healthy. If we don't, they start to wither.

Muscle mass decreases with age unless we have an active exercise routine. After the age of 30, you lose as much as 5% with each passing decade. During his lifetime, the average man will lose 30% of his muscle mass.[11] Just like you can go from a 97-pound weakling to a 200-pound gorilla, the process runs in reverse.

2.5. If we don't use our muscles, they quickly begin to shrink.

The same thing happens inside our skulls. If we don't use parts of our brain, they can begin to shrink. The process is called atrophy. It includes the loss of neurons, as well as a decrease in the number of synaptic connections between neurons.

Brain atrophy comes in two forms. The first is a *natural shrinkage* of our brains as we age. When you're in your 30s and 40s, your brain begins to shrink slightly, and the process accelerates in your 60s.[12] Each decade starting around age 40, your brain shrinks by around 5%, and this percentage increases after age 70. By the age of 85, your brain can be 25% smaller.[13]

This natural process of brain shrinkage isn't uniform. Certain brain regions shrink more and shrink quicker, and the process becomes more pronounced as we age.

Your neocortex, the brain's outer layer, which makes up more than 80% of the total mass of the brain, gets thinner as you age. Scientists have measured the average rate of shrinkage. Without even knowing your *chronological* age, they can pinpoint your brain's *biological* age using MRI anatomical scans showing how much it has shrunk.

Among the regions that shrink the most are the frontal and temporal lobes. During adolescence, these are the last parts of the brain to mature. They are the first parts of the brain to begin to shrink.[14]

The temporal lobe is located behind your ears. It processes functions like reading, writing, speaking, and understanding the meaning of words. The frontal lobe includes functions like problem-solving, impulse control, emotion regulation, memory, planning, and aspects of social interaction.

While brain shrinkage is an inevitable phenomenon with age, it's less pronounced in healthy people. There are many behaviors that support a healthy brain with minimal shrinkage. Physical exercise, meditation, a heart-healthy diet, low stress, rich social connections, continuous mental challenge, and curbing alcohol, drugs, and smoking all contribute to minimal shrinkage as we age.

The second form of brain atrophy is *accelerated shrinkage*. This leads to cognitive decline and neurodegenerative diseases like senile dementia and Alzheimer's disease. The brains of Alzheimer's patients can shrink by *as much as two-thirds.*[15]

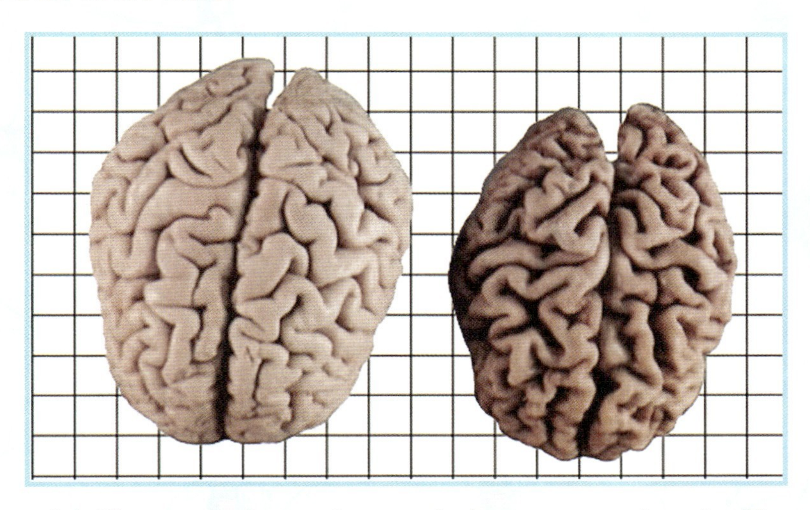

2.6. These two brains are from people the same age and gender. The Alzheimer's brain (right) has only one-third the mass of the normal brain.[16] *Note not just the overall brain size difference, but how much bigger the gaps between the folds of the Alzheimer's brain are.*

SAILING INTO THE DARKNESS

2.7. Iris Murdoch.

Iris Murdoch (1919–1999) was a philosopher and Anglo-Irish writer whose many novels were characterized by wit, intelligence, profundity, and psychologically complex plots.

Her first novel, *Under the Net,* was published in 1954 and later chosen by the American Modern Library as one of the 100 best English-language novels of the 20th century. Incredibly prolific, she wrote 25 more novels over the next 41 years. Her last novel, *Jackson's Dilemma,* was published in 1995, a year after she was diagnosed with Alzheimer's disease, the symptoms of which she at first mistook for writer's block.

In the 2001 film *Iris* depicting her struggle with Alzheimer's, Dame Iris Murdoch (she was made a Dame of the British Empire in 1987) is played by Dame Judi Dench. The movie was based on the best-selling book *Elegy for Iris* by John Bayley, the writer and English professor Murdoch married in 1956. The book bears witness to both their enduring marriage and their journey with Alzheimer's.

In an article in the *New Yorker* published the year before his book came out, Bayley writes of Alzheimer's:

> "Its victims are not always gentle: I know that. But Iris remains her old self in many ways. The power of concentration has gone, along with the ability to form coherent sentences, and to remember where she is or has been. She does not know that she has written 26 remarkable novels, as well as her books on philosophy; received honorary doctorates from major universities; become a Dame of the British Empire."[17]

Regarding the communication between them, Bayley likens it to "underwater sonar, each of us bouncing pulsations off the other and listening for an echo."[18]

Murdoch described her Alzheimer's as "sailing into the darkness."[19] In *Elegy for Iris,* Bayley says that Iris became like "a very nice 3-year-old" who needed to be fed, bathed, and changed.

At the end of the book, he gives poignant testimony to their love: "Every day, we are physically close.... She is not sailing into the dark. The voyage is over, and under the dark escort of Alzheimer's, she has arrived somewhere. So have I."[20]

Iris Murdoch died in 1999 with her husband at her bedside.

As we see from the story above, as the brain shrinks, mental and physical functioning declines. At first, brain shrinkage might show up as an innocuous symptom like forgetting names, or writer's block in Iris's case. As it loses further neurons, the brain eventually "forgets" how to balance, navigate, digest, excrete, and dress.

Alzheimer's patients have tens of thousands fewer new neurons than those of healthy controls, regardless of age.[21] The difference is especially pronounced in the dentate gyrus or DG, a part of the brain so important to SQ that we'll take a closer look at it later in this chapter. And while brain atrophy is particularly pronounced in Alzheimer's disease, it is present in many other conditions associated with cognitive decline.

Among the symptoms of the shrinking and declining brain are memory loss, difficulty in assembling strings of words, problems with physical coordination, poor judgment, mood disturbances, loss of empathy, and difficulty carrying out routine daily activities.

The big idea in this chapter is that *parts of our brain are growing, especially the parts we use, but parts are shrinking, especially those we don't.* Though we can't see what's going on inside our skulls, we can be reverting to a 97-pound weakling of a brain without the process being visible to us.

In Chapter 4, we're going to look at the experiences of people with high SQ. We'll also review the exciting research that shows that SQ grows parts of our brains and contributes to increased mass even as we age. Some of the SQ experts we'll encounter in this and other chapters have brains *a decade younger than their chronological age.*

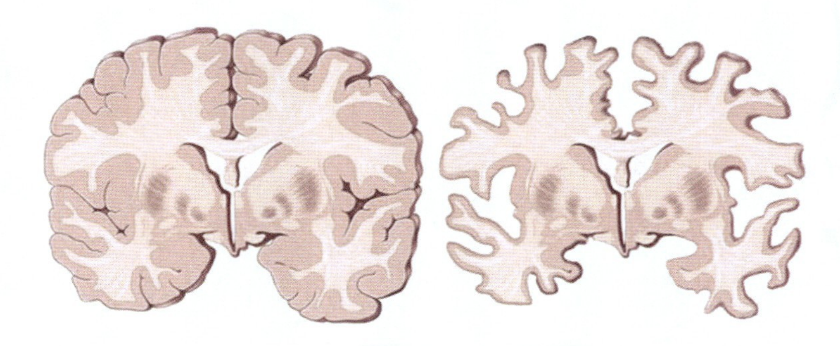

2.8. The thickness of the gray matter in the brain's cortex (the outer layer above) is a reliable measure of aging. Left: Normal cortical thickness. Right: Reduced cortical thickness.

What is it about SQ that produces growth in our brains and preserves neural mass as we age? Science is giving us fascinating pointers showing that the elevated mental states produced by SQ can stimulate positive neural plasticity.

Using the Software of Your Mind to Change the Hardware of Your Brain

The astonishing finding of MRI research is that the brains of people having the types of peak experiences characteristic of SQ can change immediately. Neuroscientist Andrew Newberg, coauthor of *How Enlightenment Changes Your Brain,* finds that the brain that comes out of a mystical experience is not the same brain that went into it. Even a single peak experience of mystical union can result in a reorganization of the brain's information flows.

When these states are induced over time, such as through daily meditation, they can change the brain quickly and radically. As SQ activates certain neural bundles to fire together, they wire together. They begin to grow. The following story is a stunning account that illustrates the speed of the process.

GRAHAM PHILLIPS DOES MEDITATION

In my book *Bliss Brain,* I describe the case of Graham Phillips, PhD. Phillips is an astrophysicist and television reporter who underwent MRI analysis before and after a two-month mindfulness meditation program. His DG, which coordinates circuits essential to emotion regulation, *grew by 22.8% in just eight weeks.*

2.9. Graham Phillips.

Take another one of those deep breaths and read the words in italics again. This brain structure that is key to regulating negative emotions became *one-fifth bigger in just two months.*

That means that Phillips had more neural tissue available to regulate negative emotion. Stimuli that might have bothered him before—like bad drivers, tedious work meetings, careless teammates, inconsiderate family members, paying taxes, misunderstood friends—were now problems he could handle easily.

Phillips reported that in just two weeks, he felt calmer, and had more tolerance for the stresses of life. After eight weeks, he had nearly a quarter more neural tissue available to handle triggering events. Other parts of his brain grew by 2% to 5%. That's a stunning change for such a short time.

Imagine if you went to the gym and lifted weights, and your arm muscles grew 22.8% bigger in eight weeks. You'd have a body like Dwayne Johnson in less than a year. Producing muscle growth that dramatic is hardly imaginable. Yet parts of your brain are performing this magic trick all the time.

Imagine if you had 22.8% more neural circuitry available to regulate anger, frustration, resentment, blame, and shame. You'd be able to dial down those negative emotions so much more effectively, paving the way for happiness and contentment. You'd enjoy a completely different life.

Phillips is just one person, but there are many studies using MRIs and collectively they measure the brains of thousands of people. Meta-analyses, which combine these studies to assess their collective results, are equally impressive. They find that brain regions associated with suffering and self-absorption shut down, while those associated with emotional self-control, compassion, and equanimity light up. Stress networks shrink, while the Enlightenment Network of Chapter 1 lights up.

Spiritual Intelligence doesn't just produce improved mood and health. As we increase our SQ, we're using our minds differently. This produces radical changes in the hardware of our brains over time, as we'll see in Chapter 6. We don't just feel better. Our brains function better. They rewire to give us a happier and more productive life.

I cannot overemphasize the importance of this scientific finding. The practices described in this book have the power to *rapidly and radically change the function and structure of your brain.* Read and use the Deepening Practices at the end of each chapter and you're likely to emerge from our journey together with a different brain. SQ isn't an abstract academic concept; it's your most powerful tool for wiring a happy, creative, and resilient brain.

IQ. EQ. & SQ

Scientists have been studying IQ for over a century. They have found that intelligence is processed primarily by the brain's frontal and parietal lobes.[22] IQ engages the brain's prefrontal cortex, especially the dorsolateral prefrontal cortex (dlPFC) and structures associated with memory, reasoning, and abstract thought. A bigger brain correlates with increased intelligence. IQ recruits all the brain's key areas, including all four lobes of the cortex, the limbic system, and the cerebellum or hindbrain.[23] People who exercise their cognitive abilities create new neurons in several of these structures.

2.10. Our cognitive abilities are sharpened by challenge at any age.

Emotional intelligence or EQ uses its own network. Daniel Goleman wrote the best-selling book *Emotional Intelligence*. His model breaks EQ down into four components: Self Awareness, Self-Management, Social Awareness, and Relationship Management. He has also collected the work of other scientists to correlate these domains with brain regions.[24]

Two of these are particularly important. The first is the amygdala. This almond-shaped structure is part of the midbrain or limbic system, which processes emotion, memory, and learning. The amygdala is triggered by emotional arousal. In *Bliss Brain,* I call it the "fire alarm of the brain" because once activated, it sends the signals that trigger the fight-flight response throughout the body. Caveman Brain doesn't have much discrimination. It fires the amygdala immediately in response to threats, real and imagined.

2.11. Prosocial emotions create empathy and caring for others.

But Goleman shows that this response is regulated by the reasoning, inhibition, and decision-making functions of the prefrontal cortex. An emotionally intelligent person can suppress the urges of Caveman Brain using this brain region. Research at the renowned Max Planck Institute in Germany demonstrates that when you make a regular practice of holding positive thoughts towards others, your brain circuitry for compassion and other "prosocial" emotions grows.[25] With practice, your EQ increases.

As we saw in Chapter 1, SQ activates its own unique set of brain circuits. The self-focused parts of the brain shut down while the Attention, Emotion Regulation, and Compassion Circuits light up. Acting together, they form the Enlightenment Network. Just as IQ and EQ can be matched with distinct parts of the brain, so can SQ. Now that the circuits for SQ have been identified, our scientific understanding of the brain's role in SQ will continue to develop.

The Dentate Gyrus

Your DG is relatively small. Think of the fork you use at dinnertime. The DJ is about the size and shape of a single prong of that fork.

Yet this tiny structure is of outsize importance. It's situated at one end of the hippocampus, your brain's memory and learning center. It functions as a gatekeeper, sifting the information going into the hippocampus and influencing which of the millions of incoming signals your hippocampus is going to process. That sorting function is one of the characteristics that makes your DG key to regulating your emotions.[26]

2.12. Hippocampus and seahorse.

The DG and hippocampus have a long history. They were first described in the 1500s by an Italian doctor named Julius Aranzi.[27] He named the hippocampus after a seahorse, because they bear a striking anatomical resemblance.

A precocious young man, Aranzi began studying medicine at the age of 19 and graduated from the University of Bologna in 1556. In a long and distinguished career as a surgeon and researcher, Aranzi treated many members of the aristocracy and published many articles on his anatomical discoveries.[28]

While neuroscientists have known about these structures for over 500 years, the functions they perform have been discovered only recently. In the 1980s it was discovered that the hippocampus was crucial to memory and learning. It plays a role in the formation of new memories and the detection of novel events, places, and stimuli.[29]

The DG analyzes and categorizes information so the hippocampus can better interpret it. To do this, it uses a widespread net of connections with other brain regions that regulate emotion. It contains neural clusters that specialize in reducing anxiety, enhancing learning, boosting memory, and suppressing fear.[30]

2.13. Neural stem cells in the C-shaped dentate gyrus.[31]

That's why when Graham Philips's DG grew by 22.8% during his eight-week meditation journey, he gained greater control over his emotions. Like a bodybuilder who had bulked up his muscles, Phillips had the neural mass in his brain to dial down anger, resentment, irritability, alarm, panic, worry, and all the other emotions that cause us pain. As his SQ increased and he routinely activated the four brain circuits we identified in Chapter 1, his DG grew bigger.

Size Matters

There are many advantages of using SQ to grow a big DG as well as other helpful brain regions. Here, in the final part of this chapter, we'll take a look at the side effects of an effective and consistent SQ practice. We'll examine research linking SQ and especially meditation to brain size and mood. We'll also face the dark side, reviewing how stress and psychological trauma shrink important regions of the brain.[32]

A key study entitled "Forever Young(er)" examined the size of various brain regions in a large sample of 100 long-term meditators. Their ages ranged from 24 to 77 years. It found that while meditators still lost brain tissue as they aged, the reduction was significantly less than that found in non-meditators. The effects weren't just found in a few regions like the DG but in many areas of the brain, including components of the Enlightenment Network.[33]

One area in which the study found growth was the corpus callosum, the bridge that allows the left half of your brain to chat with the right. Those with more tissue in this region have a bigger bridge, which enables more traffic to flow in each direction. This means we can *gather and integrate information* from many specialized regions in both the left and right hemispheres, bringing *whole-brain coordination* to bear on complex tasks.[34] In Chapter 5, we'll see how this leads to the proliferation of innovative and creative ideas in the mind.

ALBERT EINSTEIN AND THE BRAIN'S TWO HEMISPHERES

The genius of some creatives is only recognized long after they're gone. Think Vincent van Gogh, Galileo Galilei, Emily Dickinson, John Keats, Edgar Allan Poe, James Dean, and Bruce Lee.

But some rise to international fame and recognition during their lifetime. That was the fate of Albert Einstein.

When he was only 40, front-page stories about his theory of relativity had already appeared in the *New York Times,* the *Times of London,* and many other newspapers. He won a Nobel prize at the age of 42.

2.14. Albert Einstein at the time he won the Nobel Prize.

As neuroscience was coming into its own at the same time, there was intense scientific interest in how Einstein's brain worked. When

Einstein died in 1955, the on-call pathologist, Thomas Harvey, performed the autopsy.

Acting on his scientific fervor and unbeknownst to the family, Harvey removed Einstein's brain so he could study it (only later obtaining permission from Einstein's oldest son). He dissected it, photographed it, took slices that he mounted on slides, and then preserved the rest of the brain in a jar.

The brain eventually found its way to Einstein's heirs who donated it to the Mütter Medical Museum in Philadelphia, Pennsylvania.[35]

It turned out that Einstein's brain was no bigger than average. Most regions were also the same as those of the average man of his age.

However, he had a prodigious amount of neural tissue in the corpus callosum.[36] He didn't need a gigantic brain to be a genius; he only needed to have the neurons required to make the two hemispheres work in coordination.

The hippocampus, our brain's memory and learning center, is also capable of controlling function in several distant parts of the brain simultaneously.[37] Neurons running to and from the hippocampus's DG to other brain regions form a rich network of information flow, able to regulate emotion brain-wide.

When you meditate regularly, you grow key parts of your brain like the DG and corpus callosum. But that's just the start of the benefits you gain. You also increase coordination between your right and left hemispheres and between the various parts of your brain. You're not just increasing your inventory of neural hardware; you're developing the ability to use all these sections of your brain together effectively, like Albert Einstein did. This might not earn you a Nobel prize, but it's going to give you the leverage of increased wisdom and SQ every moment of each day.

Mood and Dentate Gyrus Size

Your mood and the growth of your DG go hand in hand. In one key study, researchers assigned themselves the fun job of tickling socially isolated adolescent rats. They found that the pleasurable emotions elicited by the physical stimulation increased the growth of new neurons in the DG. It also boosted the expression of genes related to health, like those that regulate feeding, cell repair, and blood pressure.[38]

The investigators summarized the large body of research with meditators and showed how consistently meditation improves both gene expression and mood. This relationship shows up at all stages of life, young and old, including people at risk of Alzheimer's disease.[39]

The effect of a consistent practice shows up quickly. Even *13 minutes of meditation a day* results in better mood, attention, memory, and emotion regulation.[40] When you keep it up for eight weeks or more, the effects extend further, boosting your ability to focus and concentrate, sort the trivial from the important, and retain key memories. A review of all the current studies of mindfulness found that it consistently improves cognitive ability.[41]

When the human mind is filled with positive emotions and elevated moods, like those generated by laughter and meditation, the DG grows.[42] Those new neurons in the DG improve your ability to control anxiety, increase learning and memory, and enhance your ability to master your fears.[43]

By now the exciting message of all this research is clear. When we meditate, we initiate a virtuous cycle. We improve our mood while simultaneously increasing the size of the DG. That super-sized DG is then better able to control our fears, screen out negative incoming signals, reduce our anxiety, and coordinate multiple brain regions for positive emotion.

As well as improving our health body-wide, the process can boost our memory, increase our capacity for learning, improve our cognitive abilities, and reduce our risk for Alzheimer's and other forms of dementia. This in turn improves our mood, and the cycle continues.

Shrinkage of the Dentate Gyrus

However, the reverse is also true. Stress and negative emotion can also affect the DG, shrinking it and inhibiting the growth that comes with positive emotion. The dark side is a vicious cycle of diminished mood and neural shrinkage.

Both chronic and acute stress change the length and density of neural spindles and the number of connections they form with adjacent neurons. The key brain regions affected by this are the hippocampus, prefrontal cortex, and amygdala.[44]

Chronic stress actually inhibits the growth of neurons in the DG. At the same time, it produces increased growth in the brain's "fire alarm" amygdala. This adverse neural plasticity is significantly associated with increased anxiety, aggression, impulsiveness, poor decisions, abusive relationships, and low self-esteem.

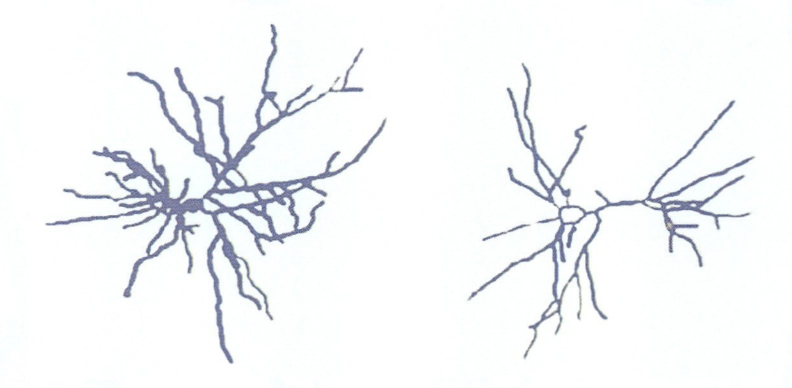

2.15. Stress is associated with a reduction of the number of neural connections in the DG, hippocampus, and prefrontal cortex. Left: normal density. Right: reduced density in stressed brain.[47]

Children raised by abusive mothers have larger amygdalas. Maternal depression also produces increased amygdala volume in children. Tripping the fire alarm repeatedly tells the body that this part of the brain is vital to survival, so it invests additional resources into growing it.[45]

The bad news doesn't stop there. The executive center of the brain, the prefrontal cortex, starts shrinking, diminishing our ability to regulate those negative emotions.[46]

The Nucleus Accumbens

Another part of the brain affected by stress is the nucleus accumbens. This structure is a key part of the brain's reward system. It's active in cravings and addictions, whether they're to chocolate, sex, shopping, fentanyl, TV, gambling, or whiskey.

The nucleus accumbens plays a role in reinforcing behaviors and in the stimulus-reward cycle. It manages the progression from simply *experiencing something as rewarding* to compulsively *seeking that thing out* as part of an addictive pattern. It's activated when we even anticipate experiencing something pleasurable. The nucleus accumbens might light up when a vaper just glances at a vaping pen or even thinks about a vape. It is highly activated by addictive drugs like fentanyl.

In master meditators like Tibetan monks who've spent 10,000 hours or more meditating, the nucleus accumbens shrinks. With the four circuits of the Enlightenment Network active, desire for anything other than the ecstasy of oneness with the infinite vanishes.

But in addicts, the nucleus accumbens grows. It becomes more sensitive to addictive stimuli with each fix.

Stress Alters Key Brain Regions

The nucleus accumbens is also affected by stress. Stress changes the structure of neurons in the nucleus accumbens in a way that boosts addictive cravings, making it harder and harder to resist that fix.[48]

Another brain structure that shrinks in those master meditators is the amygdala. They simply aren't activating the fire alarm regularly, so the body recognizes that it doesn't need all those excess neurons. Even in novice meditators, an eight-week practice can produce shrinkage in the amygdala.[49]

That's how fast meditation can remodel your brain.

When children are stressed early in life, which often happens when the adults around them aren't capable of regulating their own emotions, the effects on the brain show up decades later. Early life stress inhibits the formation of new neurons in the adult DG.[50]

Again, the process of brain remodeling starts astoundingly quickly. The DG starts to change when animals are exposed to stress for *only two hours*. The software of emotion acts as an epigenetic signal that changes the hardware of the brain.[51]

The evidence for the association between stress and shrinkage of neural mass in key brain regions like the DG is substantial. This traps people in *a downward spiral of mood and physiology*. We get stressed and our emotions suppress the growth of emotion-regulation neurons in our DG.

We now have less neural capacity to handle stress, so we become stressed more easily. The increased stress further impairs the structure of the DG. As the cycle continues, we become more and more miserable.

Yet research shows that meditation arrests the downward spiral. As we breathe consciously, become mindful, and activate SQ, we stop the slide. We begin to build neurons in the DG and other components of the Enlightenment Network. In *as little as 15 minutes a day, in just four weeks,* our brains improve. Positive brain changes persist over time; even two weeks after meditating, the brain's attention and self-transcendence circuits are still more highly engaged.[52]

With extended practice, we train our brains into an upward spiral. Greater SQ produces positive emotion and bigger mass in key parts of our brains. We've now redirected ourselves toward a lifetime of thriving. In Chapter 4, we'll see the heights of ecstasy to which this can take us.

Repair of the Dentate Gyrus

What happens in the brain as we meditate daily and throw ourselves into the joyful journey to SQ? We enter the upward spiral, engage the Enlightenment Network, and start attenuating negative emotion.

Old fears and anxieties can be overwritten by the DG. In a study using mice, researchers found that the less scared the mice were, the more neurons in the DG activated. The mice became more fearful when the investigators reduced the excitability of DG neurons, and less fearful when they enhanced excitability.[53]

In a similar study using rats, the researchers created conditions that simulated depression in humans. Some developed "learned helplessness," becoming so depressed they simply gave up. But when the investigators stimulated the DG, their behavior came to resemble that of rats with active coping strategies.[54]

LIFTING THE WEIGHT OF AN ELEPHANT OFF MY SHOULDERS

Trixie South battled severe clinical depression for over a decade. Her condition was characterized by persistent feelings of sadness, hopelessness, and a loss of interest in activities she once enjoyed. It weighed her down so much she called it "the elephant sitting on my shoulders."

After being diagnosed by her doctor with major depressive disorder, she tried various medications and therapies. But her symptoms persisted, significantly impairing her quality of life. Her condition led to frequent absences from work, strained relationships, and a profound sense of isolation.

In Trixie's words: "Depression feels like being trapped in a dark, endless tunnel with no way out. Every day is a struggle to find the energy to even get out of bed. There's this constant, overwhelming sense of hopelessness and despair that makes it hard to see any possibility of a better future."

Trixie's therapist had been reading an ancient scripture called the *Yoga Sutras* of Patanjali, one of the classic texts on enlightenment. A class on Patanjali's meditation techniques was available at her local college, and after her therapist recommended it, Trixie enrolled.

She learned mindful breathing, scanning her body to identify areas of tension, basic yoga asanas, proper meditation posture, and loving-kindness meditation. Trixie committed to a daily meditation practice, dedicating 30 minutes each morning to these exercises.

Within the first two weeks, Trixie noticed subtle changes in her mental state. She began to experience brief moments of peace and clarity, which provided a stark contrast to her usual turmoil.

Encouraged by these glimpses of relief, she continued her practice with increased dedication.

Over the following months, Trixie's transformation became more pronounced. Her mood improved and she started to feel a renewed sense of purpose and connection. The once overwhelming feelings of sadness and hopelessness gradually diminished, replaced by a growing sense of wellbeing and resilience.

She observed, "My practice has been my lifeline. It didn't happen overnight, but slowly, the fog of depression began to lift, and I started to see glimpses of hope and happiness. I feel like I'm finally living again."

2.16. The joy of rollerblading.

After six months of consistent meditation practice, Trixie's life had improved dramatically. She returned to work with a newfound energy and enthusiasm, rekindled relationships with family and friends, and experimented with new hobbies. She became passionate about rollerblading.

Her therapist noted significant improvements in her clinical assessments, while Trixie felt more balanced and in control of her emotions. Meditation had become a cornerstone of her daily routine, providing her with a powerful tool to manage her depression and cultivate a more fulfilling life.

Trixie put it this way: "I used to wake up every morning dreading the day ahead. But nowadays I start by finding joy in the small things. The elephant has finally been lifted off my shoulders."

People in Caveman Brain mode are not great thinkers and research shows they have lowered cognitive ability. They're anxious and depressed, ruminating on the tiger that almost ate them yesterday. They alternate this with worrying about the hypothetical tiger that might eat them tomorrow. When anxiety is dialed down by SQ, the Enlightenment Network activates and cognitive processing improves.

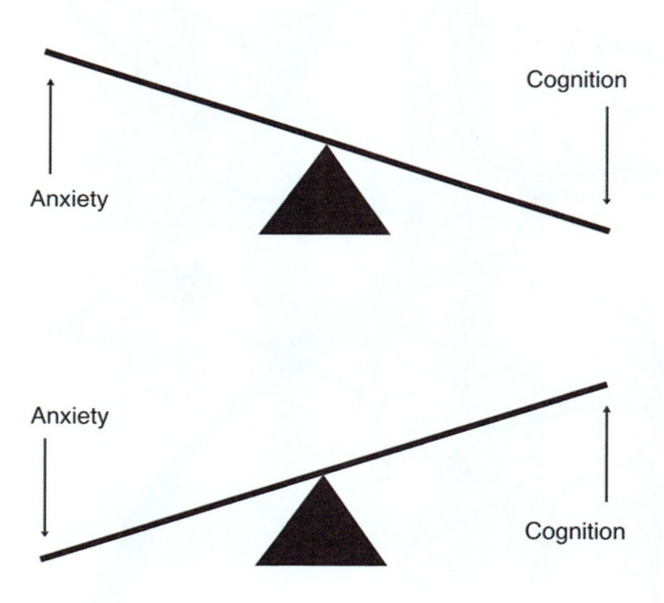

2.17. Anxiety and cognition have an inverse relationship; when one goes up, the other goes down, and vice versa.

The DG is associated with both of these simultaneously. It assists us with both attenuating anxiety and boosting cognition. We can think of anxiety and cognition as two sides of a seesaw or teeter-totter. When one goes up, the other goes down, and vice versa. A Dwayne Johnson–sized DG gives us the benefit of both improvements at the same time.[55]

Meditation Improves Alzheimer's Brains

We'll close this chapter by looking at the changes that meditation produces in the brains of Alzheimer's patients. Perhaps someone in your family or social circle is affected by Alzheimer's or another form of dementia.

That's most of us. A survey found that 56% of US adults said they had a family member or friend who had Alzheimer's.[56] This disease takes a huge human toll on loved ones. More than 16 million Americans provide more than *17 billion hours of unpaid care* for family and friends with Alzheimer's disease and related dementias each and every year.[57]

Perhaps you're not one of those, but you're concerned about cognitive decline as you get older. You've watched people you love "lose it" as they age, and worry that you might wind up like them one day. You might have taken a genetic test from one of the many companies that supplies them affordably and discovered you have a gene like the APOE4 gene that predisposes you to Alzheimer's. For any and all of these reasons, it's worth exploring what SQ in general and meditation in particular can do to address these conditions.

A key study examined the effects of long-term meditation on brain structures of patients with mild cognitive impairment and Alzheimer's disease.[58] A control group received coloring books to fill out. Both groups received MRI scans at the start of the study and after six months of practice.

In the meditation group, participants completed four personal training sessions over the course of two weeks. The meditative procedures came from classical scriptures that we'll explore further in Chapter 4. These included the *Yoga Sutras of Patanjali*, the same source Trixie used, and the *Path of Purification,* a condensation of Buddhist philosophy and meditation techniques.

Alzheimer's patients learned a combination of four procedures, one at each training session. The first of these included *full-body relaxation,* systematically surveying the entire body beginning with the feet and traveling to the head, consciously relaxing all muscular and nervous tension.

This was followed by the other three techniques: witnessing the *breath* flowing in and out, the *contemplation* of one of life's most peaceful moments, and noticing the progression of your thoughts from a detached *witness perspective.*

These four components are closely related to the techniques employed in EcoMeditation. To the body scan performed by study participants at the start of each day's meditation, EcoMeditation adds a stress-release protocol based on acupressure. It then layers in a structured breathing rhythm.

This is followed by a technique from the Princeton Brain Lab that evokes alpha brain waves, leading to a feeling of "oceanic bliss." The fourth technique, adopting the witness perspective, is accomplished by a simple method that dials down activity in the brain's narcissistic Default Mode Network. In Chapter 4, we'll learn more about the mystical experience this produces.

In the Alzheimer's study, the trainer also provided patients with step-by-step instruction in how to induce a deep meditative state. After having their questions answered, patients received a detailed written handout and a guided audio CD. For the next six months, they practiced at home for 30 minutes a day, supported by the research team. They then received their second MRI scan.

Alzheimer's Patients Grow the Enlightenment Network

Needless to say, coloring books produced no changes in the brains of the control group. But in the meditation group, widespread and significant changes were found in several areas of the brain, including components of the Enlightenment Network.

Their scans showed increased gray matter volume and cortical thickness in areas related to attention, executive function, and memory such as the prefrontal cortex. It also found growth in regions involved in emotion regulation and the inhibition of conditioned responses. Earlier work had found thinning in some of these same areas in Alzheimer's patients.[59]

The meditators had decreased cortical thickness and gray matter volume in regions toward the back of the brain such as the posterior cingulate cortex. That's one of the two poles of the Default Mode Network, the network active in self-absorption and unhappiness. Meditation was also linked to increased volume in the right thalamus, a structure that serves as a gatekeeper for body signals.

Those body signals like pain and hunger can distract us from our meditative practice, and when the thalamus is active, it screens these out so we can maintain inner peace.[60]

The authors of the Alzheimers's study concluded, "An increase in gray matter predominantly over the prefrontal cortex and a decrease more posteriorly suggest a shift toward increasing top-down control in the meditators. More importantly, it suggests that long-term daily meditation may be able to slow-down the neurodegenerative process in crucial areas of the brain."

Spiritual Intelligence Counters Cognitive Decline

There is an entire body of research that supports the conclusions of this key Alzheimer's study. A systematic review of all published research on the topic of religion and spirituality in cognitive function found that *82% of the research articles reported positive associations with brain health.*[61] Not only is SQ protective against cognitive decline, but the reverse is true: Lower spiritual wellbeing is associated with mild cognitive impairment and early dementia.[62]

Andrew Newberg, coauthor of *How Enlightenment Changes Your Brain,* believes that the evidence is so compelling that "spiritual fitness" should be considered a key component of any cognitive health program.[63]

MOM'S DRIVING AND ALZHEIMER'S

By Irene McKenna

My mother had never been a great driver, but she got by. However, after her 60th birthday, we noticed that it was getting progressively worse. By the time Mum got a diagnosis of Alzheimer's at 65, she was driving like a maniac. It reached a point where I flatly refused to go anywhere with her unless I could take the wheel.

Our doctor prescribed a new Alzheimer's drug, but my brother David is a research chemist and took a deep dive into the research behind the drug. He found that while it was effective at removing the sticky plaques that are found in the brains of many patients, it did not improve cognitive ability. And the side effects were horrible, with more people harmed than helped by taking it.

We eventually got a referral to Sally, a naturopath who helped Mum clean up her diet and start taking helpful herbal supplements. Sally also recommended she begin meditating and gave her a couple of apps to make it easy.

Around that time I lost my job, so I decided to move in with Mum. It was a challenge, but we enjoyed cooking together. I could also remind her to take her supplements and medications. And it was a good excuse for me to take the wheel whenever we had to drive.

We began meditating together each morning. The guided meditation tracks made it easy to follow. After a cuppa, we'd sit together in the parlor for about 30 minutes at the start of each day. I could see the improvement in Mum's condition after only a few weeks. She was noticeably calmer and less anxious about life. David, who'd been skeptical about this "woo woo" stuff, stopped scoffing at it.

I kept on staying with Mum even after I got a part-time job. We'd begun to look forward to our mornings together so much that our meditation sessions stretched to 45 then 60 minutes. She was now less forgetful and could follow the thread of a conversation with David and me.

Then one fateful day she got behind the wheel of her car again. I had a hard time keeping quiet and I squirmed as she pulled into the lane.

But the old Mum was back! Her driving still wasn't perfect, but it was no worse than it had been before the Alzheimer's diagnosis. I started to cry silently because I felt I had got my mother back after all those stressful years.

2.18. Mother and daughter meditating together.

The next test will be when I move out. But Mum's connecting with her friends again, making shopping lists, remembering recipes, driving around buying the stuff she needs, talking lucidly to me and David regularly, and seems to be in generally good shape.

It's like there was a pre-meditation version of Mum and a post-meditation version. David and I were both skeptical when Sally recommended meditation, but now even he has begun to join us some mornings.

Behind all the statistics in the Alzheimer's studies cited here, there are real people. They've seen their loved ones improve, just like Irene's mother did. There are now many people who, like her skeptical brother David, have seen the changes meditation can produce.

Even if you're in your 20s or 30s, it's worth embracing a consistent meditation practice now. While the symptoms of cognitive decline typically show up in our 60s, imaging methods like SPECT scans show that *adverse brain changes are evident up to 30 years before* we develop externally observable symptoms like forgetfulness.[64] Alzheimer's proteins are also detectable in the blood decades before the disease overwhelms the brain's regenerative capabilities.[65]

Being proactive and starting an early life program for brain health gives you both the immediate benefits—like a sharper mind, more focused attention, and reduced anxiety—and the comfort of knowing you're predisposing yourself for the best possible Golden Years.

Spiritual Intelligence Shapes the Structure and Function of Your Brain

The initial studies on brain regions used Tibetan monks as their subjects. These master meditators have at least 10,000 hours of practice. Some have over 50,000 hours. They're also using traditional techniques refined over centuries, like those described in the *Yoga Sutras of Patanjali,* which we'll explore in Chapter 4. This produces such pronounced changes in the brain that they are easy to detect.

Remember, in the early 1990s, when researchers first studied monks' brains, they had MRIs of much lower resolution than we have now. Today, resolution has increased to the point where we can study the development of a single neuron. But the magnitude of brain change produced by meditation is so great that even those first primitive MRI scans were able to detect it. Today, new MRI machines are showing us that brain change is much faster and more extensive than we first realized. SQ is producing younger, healthier, and happier brains, remodeling an extensive network of regions, as we'll see in Chapter 6.

A BRAIN NINE YEARS YOUNGER

Neuroscientist Richard Davidson has been studying meditators at his neuroimaging lab at the University of Wisconsin since the 1980s. In one study, he tested 21 adepts. The most experienced of these was a Tibetan monk called Mingyur Rinpoche.

2.19. Mingyur Rinpoche.

Even when Mingyur was a toddler, his predisposition to a devotional life was apparent. When mother asked what game he'd like to play, his favorite was pretending to go into a cave to meditate like a monk. As a result, by the time he was 42 years old, he'd clocked over 62,000 hours of meditation.

When Davidson first conducted an MRI on Mingyur, he gave the monk a number of exercises to complete in order to determine which brain regions were engaged by each one.

When Mingyur was given the cue to engage compassion, Davidson and the other researchers in the lab's control room were stunned. The level of activity in Mingyur's Empathy Circuit rose by 700%.

"Such an extreme increase befuddles science," wrote Davidson.[66] On average, the adepts had 25 times the amount of gamma, a brain wave associated with happiness, integration, and creativity, when compared to a control group.

This slowed down brain aging dramatically. The researchers measured the thickness of Mingyur's cortex, a standard measure of the brain's biological age. They found it to be a staggering *nine years younger* than his chronological age.

Take a breath and think about that research finding for a moment. Wouldn't getting twice as happy as you are today be great? Perhaps three times? Can you even imagine being *25 times as happy*?

Increasing Happiness 25X

Most of us cannot even imagine such a possibility. Yet that's exactly what science shows us is possible when we develop SQ. The side effect of that much happiness is a young brain. Imagine having your brain be nine years younger than your chronological age. Your mental activity would be much sharper than that of your peers all the way to the end of your life. That's the promise of SQ.

The other thrilling new development is comparing the brains of these master meditators with those of novices. Few of us are ever going to match feats like Mingyur's 62,000 hours of meditation. Unless you're a monk, you're unlikely to clock even 10,000 hours in your lifetime.

What the research outlined in this chapter demonstrates—along with hundreds of other studies that support the results—is that *even a single meditation session* produces changes in your brain.[67] In novices starting to use a scientifically proven method like EcoMeditation, just a few minutes each day for a few weeks shrinks the circuits of suffering and grows those of happiness. When you turn on the Enlightenment Network and develop SQ, radical shifts take place in the structure and function of your brain.

In the next chapter, we'll look at the implications of where these changes take us. There's a specific point in brain development where our whole lives change, quickly and radically.

In Chapter 4, we'll look at the characteristics of the awakened experience and discover the SQ of the mystics. I'll also outline the specific meditation methods that produce these changes and how you can easily layer these into whatever style of meditation you most prefer.

Make sure you explore the Extended Play resources at the end of this and every chapter. These contain audio tracks you can use in your own meditation journey, videos and websites for further exploration, and updates as new research is published.

Of all the journeys a human being can take, the one you and I are on now is the most exciting. SQ takes us to the edges of consciousness to explore the biggest questions in life. There we discover happiness, love, and meaning beyond anything we might have believed humanly possible.

Deepening Practices

The Deepening Practices for this chapter include:

- **Journaling Exercise 1:** What would your life look like if you were twice as happy? Write as many bullet-point details as you can imagine. Creating a vision is the start of manifestation.
- **Journaling Exercise 2:** What stresses you out on a consistent basis? Make a list in your journal of those areas of your life where you could most use more emotion regulation. You're likely to see these stress patterns changing as you practice the resources in this book.
- **Journaling 3:** What activities could you do to enhance your personal brain health? Meditation is the most obvious one, and there are others that challenge your cognitive abilities. Make a list of at least three.
- **Dawson Guided EcoMeditation:** Activating Bliss Brain

Get the meditation track at: SpiritualIntelligence.info/2.

Extended Play Resources

The Extended Play version of this chapter includes:

- One minute video of an Alzheimer's brain being one-third smaller than a normal brain
- Podcast: Your Brain on Meditation: Andrew Newberg and Dawson Church in Conversation
- Teachings of Minyur Rinpoche

Get the extended play resources at: SpiritualIntelligence.info/2.

Updates to This Chapter

As new studies are published, this chapter is regularly updated. Get the most recent version at: SpiritualIntelligence.info/2.

Chapter 3

The Tipping Point

BENCH PRESSING 200 POUNDS

"Yes I can, yes I can," I chanted inside my head.

I was straining to bench press 200 pounds (91 kg).

For a couple of years, I had been building up to this moment. I'd built up the muscle fibers in my skinny arms to lift 150 pounds and every few weeks I would add an additional 5.

But it was slow going. Sometimes I would plateau for weeks, unable to lift the next 5 pounds. Other times I would improve by 10 pounds in a single week.

Eventually, after a couple of years of effort, I reached 195 pounds. After a couple more weeks of practice, I added the final 5.

"Yes I can, yes I can," I chanted as I pushed with all my might.

But I couldn't. I simply didn't have quite enough muscle fiber in my pectorals, deltoids, triceps, and other muscle groups to accomplish the feat.

3.1. Bench presses require good form and consistent practice to build up muscle mass.

I knew that mindset and belief was all-important. I had been visualizing my goal clearly all along. I affirmed that I could meet it. I believed 100%.

But I just could not get above that 195-pound threshold, despite all my positivity, belief, and affirmation, because I simply did not possess the muscle mass to accomplish the goal.

Then one bright summer day, without thinking too hard, I casually added the final 5-pound weight to the bar. I lay down, took a deep breath, and bench pressed 200 pounds.

I'd reached the tipping point.[1] I had just enough muscle fibers in my arms and chest to be able to lift that much weight.

The next day, it felt easy. A week later, I added another weight, bench pressing 205 pounds (93 Kg).

The lesson I learned here was that mindset, belief, and all the invisible qualities of the mind are not enough. They can, and do, create massive change in the tangible world, as I demonstrate in my book *Mind to Matter*.[2]

But you also need the physical ingredients to accomplish a goal. If you're baking a cake and the recipe calls for flour, water, milk, and egg, you can't proceed if you lack flour. No amount of intending, affirming, believing, and visualizing is going to make up for the absence of needed physical ingredients. No flour, no cake. Period.

What are the physical ingredients required for Spiritual Intelligence? In Chapter 4, we're going to discover the five characteristics common to people having mystical experiences. In chapter 5, we'll discover the second route into Bliss Brain, through "flow" states.[3] Both of these build neural tissue in the Enlightenment Network and facilitate SQ. Those neurons are the *physical substrate of SQ*.

In this chapter, however, we're going to explore the research showing that Spiritual Intelligence *isn't just spiritual*. It's *material*, and without the right physical ingredients, it can't be sustained.

We need to build up a critical mass of neural tissue to flip into SQ. At some later point, sustaining SQ might be effortless. But initially, we practice. We might work out for years with little apparent improvement, the way I bench pressed on my way to 200 pounds. Then one bright day, we reach the tipping point. We're in Bliss Brain and we can get there the following day and the next. Our patient practice has gradually *built enough neurons in the Enlightenment Network to reach the tipping point.* Once we've passed the tipping point, what was difficult before becomes effortless.

From States to Traits

Spontaneous experiences of SQ lead to peak inner states. When we're in a concert hall carried into rapture by the music of the band, or hugging a saint, or dancing in the light of a full moon, or gazing into the eyes of a lover, we enter the awakened state for a while.

We can't sustain these states for long periods of time. They were triggered by an *outside* influence and it takes that stimulus to awaken them within us.

But once the tipping point is reached, they become an *inner* reality and we can sustain them without an *outer* stimulus. We've built up our neural bundles through lots of time firing in that state and they're large enough to be a reliable conduit for the information flows of SQ.

At that point, this experience changes from being a *state* to a *trait.*[4] A state is temporary, while a trait is enduring. We've all experienced the state of happiness and know just how wonderful it is. But that's entirely different from *being a happy person.* Such a person has a happy disposition, a permanent, hardwired *trait* that endures through good times and bad. Their happiness is not *dependent* on what's happening around them; it's an *independent* function of the neural structure of their brain.

One of the most famous stories in history of a person reaching the tipping point is that of Gautama, the Buddha. He had a dramatic overnight enlightenment experience and never went back to his old state. He'd made enlightenment a trait. However, this "sudden" event was the fruit of many years of practice. Like a weight lifter adding the last muscle fiber needed to lift a heavy burden, that night Gautama added the last neuron.

GAUTAMA'S JOURNEY TO THE TIPPING POINT

3.2. Traditional representation of Buddha after his enlightenment.

Prince Siddhartha, the future Buddha, applied all his energy to the search for enlightenment. He spent six years studying with the two monks most esteemed for their learning. He practiced every technique to perfection. He was the best possible student, the holiest of the holy.

Siddhartha's ascetic practices were so extreme that his body wasted away. One day a passing girl named Sujata saw his emaciated form and thought he was a ghost. She fed him some broth to bring back his strength. But despite the fervor of his quest for SQ, he didn't feel he was making much internal progress and at times he despaired of breaking through.

One evening he sat with his back to a bodhi tree and began his meditation practice. But no sooner had he settled himself than a demon appeared to distract him.

This was no ordinary demon; it was Devaputra Mara, the king of the demons. *Mara* is a Sanskrit word meaning "demon" and refers to anything that obstructs the attainment of enlightenment.

Mara tried to disturb Siddhartha's single-minded focus. He knew that if he could disrupt Siddhartha's concentration, he would derail his path to enlightenment. Mara brought his entire gang and they enthusiastically tried to break Siddhartha's focus. Some demons threw arrows at him, others shot fire, and yet others hurled boulders. When these failed to distract the prince, they even picked up entire mountains and flung them at him.

Despite these afflictions, Siddhartha's concentration remained fixed on love. It turned the missiles into a rain of flowers.

Mara then changed tactics and became a ringmaster, orchestrating the ultimate Miss Universe contest. He conjured up a variety of beautiful women of all shapes, sizes, and colors—black, yellow, red, white, short, tall, round, skinny—offering Siddhartha any combination of pleasures.

Siddhartha's focus only deepened. He continued to meditate until dawn, and in this state of perfect concentration, he permanently removed the veil separating his local consciousness from universal nonlocal consciousness, becoming a perfectly enlightened being.

When he emerged from meditation the following day, everyone knew he had changed. His fellow students were awed. They asked, "Are you God?"

He replied, "No. I am awake."

Siddhartha's previous six years of focused effort had developed his SQ bit by bit, even though nothing dramatic seemed to be happening on the outside. He brought all that intention and focus to bear that night under the fig tree and reached the tipping point.

That's what a tipping point experience looks like. Some are as dramatic as Siddhartha's, while others are more subdued. What they have in common is that, after building up neural bundles quietly in their brains for years or decades, these people have enough cellular mass to produce a breakthrough. At the tipping point, a state becomes a trait and the brain is permanently changed.

Our brains are used to perceiving gradual change but are often surprised by rapid change. Watch a pot of water on the stove and nothing seems to be happening for a long time. Then suddenly it boils. Blow up a balloon and it gets slowly bigger. Suddenly it pops, startling us. We're watching a gentle rain fall from the clouds above us and suddenly there's a bolt of lightning. The lightning is scary in a way the rain is not. There are many examples in life and nature of sudden and dramatic tipping points.

The Big Mistake Made by Religion and Psychology

There is a profound misunderstanding woven into the fabric of spirituality and psychology. They often focus on bringing us into *states,* rather than fostering *traits.*

Traditional religious practices emphasize experiencing the union of human and divine. We revere those who have this mystical experience, like Buddha, Jesus, Mary, Rumi, and Gandhi. As we'll see in Chapter 4, though the world's great religions look very different from each other on the outside, at their core they're all based on the same mystical experience.

3.3. Psychotherapy session.

Psychology focuses on relieving dysphoric states such as anxiety and depression, and bringing people to a baseline in which their day-to-day experience is not haunted by these debilitating conditions.

These goals of psychology and religion are laudable. Every human being would rather be happy than miserable, and elevated states of bliss and oneness are certainly part of the mystical experience conferred by SQ.

What both psychology and religion miss is an understanding of the neurological nature of ecstasy. You need to have *the neural circuits of your brain* developed to the point where you've made happiness a *trait*. If it's still a *state,* it's dependent for its continuance on outside circumstances. No flour, no cake.

But once you've grown the neural pathways of the Enlightenment Network to a great enough size and reached the tipping point, ecstasy becomes a trait. No adverse circumstance can take it away from you.

Building the Hardware to Run the Apps

Think about your smartphone. It's a piece of computer hardware so sophisticated that this tiny device is more powerful than all the room-sized supercomputers that took the first man to the moon.

Your phone runs apps. These are pieces of software code that allow you to accomplish a huge range of functions, from keeping track of your schedule to watching a movie to finding a friend to managing your money to tracking your exercise routine.

To run the apps, you need the hardware of your phone. If you don't have the phone, all the apps still exist, but you can't access them. These pieces of software code make your life much easier and more convenient, but you need the hardware of your phone to run them.

Having the hardware is non-negotiable. No phone, no apps.

To run the app of happiness, you need the hardware of *precise neural circuits in your brain.* No hardware of circuits, no software of happiness. There are many studies that demonstrate the dependence of apps like gratitude, awe, optimism, compassion, and joy on specific neural networks in your brain. You build these neurons one by one through temporary states. Once you reach the tipping point, they're big enough to result in permanent traits. Here are a few examples.

Happiness is associated with a decrease in connectivity between your dorsolateral prefrontal cortex, part of the Emotion Regulation Circuit we examined in Chapter 1, and the Default Mode Network, which runs the "suffering self" app.[5] That means that the self-absorbed whining inner grouch of the Default Mode Network exerts less of an influence on the "smart parts" in the prefrontal cortex.

Happiness is also linked to activation of the insula, central to the Empathy Circuit, and a brain region central to many positive emotions like gratitude, awe, joy, and compassion.[6] The Default Mode Network is relatively quiet in happy people, showing less functional connectivity with other regions of the brain and thus less of a tendency to allow its rumination, catastrophization, and self-absorption to take over other areas.[7] When this piece of hardware is switched off, it takes the apps of many negative emotions down with it.

3.4. Once you've built the trait of happiness into your brain, bliss becomes your baseline.

On the flip side, adverse brain changes often precede adverse behavioral changes. A large scale study of cognitive decline was conducted at Sweden's prestigious Lund University. It examined the physiology of 356 people over the age of 65.[8] The researchers selected participants who had no symptoms of cognitive decline at the start of the project. They tested their levels of anxiety, apathy, and overall cognitive function as well as measuring the concentrations of Alzheimer's plaques in their cerebrospinal fluid.

Over the next eight years, they followed up every six months to see which patients began to show psychological signs of Alzheimer's and other forms of dementia. They found a clear link between elevated plaques at the start of the study and a later rise in apathy and anxiety—early signs of cognitive decline.

Maurits Johansson, physician and lead author of the study, explains: "Alzheimer's disease affects large parts of the brain, including the regions that control our emotional life. Our study shows that psychiatric symptoms, just like cognitive symptoms, occur mainly as a *direct consequence of the underlying changes to the brain,* due to increased levels of amyloid beta." The hardware of brain change is activating the app of cognitive decline.

As we saw in Chapter 2, the amygdalas of people with high SQ begin to shrink, and in masters like Tibetan monks they actually atrophy, because they're not activating the brain's "fire alarm" repeatedly. When the circuits of Caveman Brain are turned off, there's no hardware on which to run the app of stress.

The absence of emotional activation doesn't turn these monks into Mr. Spock of *Star Trek,* cold and unemotional. When they're confronted with an actual threat, their amygdalas still turn on. However, when exposed to positive emotional information, their amygdalas are more engaged than those of unhappy people.[9] The trait of optimism is also associated with greater connectivity between the ventromedial PFC, part of the Emotion Regulation Circuit, and other emotion processing structures in the brain.[10] This hardware runs the app of optimism.

These brain structures are particularly well-developed in meditators and they lead to an enhanced ability to *dial down negative emotion* while simultaneously *escalating good feelings.* This kind of brain has embedded the habit of happiness as a neurological trait, not just a passing emotional state. It has built the hardware to run the positive apps of joy, gratitude, awe, optimism, and compassion.

The Awakened Brain

Lisa Miller is a clinical psychologist and researcher. In her inspiring book *The Awakened Brain,* she outlines the interplay between spirituality and various brain regions, presenting compelling evidence that spirituality is deeply rooted in our neurobiology. Miller's groundbreaking MRI research highlights specific brain regions that are particularly active during spiritual experiences, showing that our innate capacity for SQ has a biological basis.[11]

One of the areas she has discovered to be key is the posterior superior parietal lobe, a region implicated in the perception of oneself in relation to the spatial environment. It's part of both the Attention Circuit and the Emotion Regulation Circuit we learned about in Chapter 1.

This area, according to Miller, plays a crucial role in spiritual experiences by helping individuals feel a sense of *connection to something greater than themselves.* The sense of interaction with universal consciousness is the core of SQ.

Another significant region identified by Miller's research team is the prefrontal cortex, particularly its association with contemplative practices like meditation and prayer. The prefrontal cortex is involved in attention, focus, and decision-making processes, and structures within it are part of all four circuits of the Enlightenment Network.

As we grow the hardware in these regions, our capacity for empathy, compassion, and ethical decision-making is enhanced. This fosters a deep sense of social connectedness and moral responsibility toward others.

Depression and Spirituality Are Two Sides of the Same Coin

Miller emphasizes the role of the limbic system, in which elements of the Empathy Circuit and Emotion Regulation Circuit reside. She finds that spiritual experiences can modulate the activity in this system, leading to heightened feelings of peace, profound joy, and a deep sense of meaning in life, characteristics of the mystical experience we'll cover in Chapter 4.

Miller has done a great deal of research with teenagers, as well as with adults.[12-13] She's identified an inverse relationship between spirituality and depression. Only 15% of teens describe themselves as spiritual. But that 15% had lower rates of anxiety and depression. Teens who don't have a spiritual life have 10 times the risk of sociopathy.

Both teenagers and adults who describe themselves as "spiritual" are less prone to depression, while those who are depressed are less likely to identify as spiritual. One of Miller's key insights is that depression and spirituality are "two sides of the same coin." One of the Extended Play resources at the end of this chapter is a lively dialog between Lisa Miller and me.

3.5. Depression and spirituality are two sides of the same coin.

Here again we see the interplay of mood and brain. Develop the brain structures of SQ and the outcome shows up in a happier mood. Cultivate positive emotion and we grow the brain structures of the Enlightenment Network. Lisa Miller observes, "Spiritual awakening depends more on the deliberate use of our inner life than on our relative endowment from biology."[14]

Like the day when my muscles had finally grown big enough to bench press 200 pounds, a person with a trait of SQ has bulked up their Enlightenment Network to the point where Bliss Brain is their baseline experience. They've also shrunk the structures that make them unhappy and decreased the chatter between these and the parts of the brain that build one's sense of identity.

When you begin viewing any inner state this way, as a set of neurons you need to amplify in order to have the desired experience, you approach it as a *neurological challenge*. You just have to *fire those neural bundles often enough and strongly enough* to make them grow. Once they reach the tipping point, you own that trait on a permanent basis. You've built the hardware to run the desired app.

The Neurological Basis of Spiritual Intelligence

This *neurological view of personality development* has profound implications. When seeking to develop SQ, you can discard most of the mythology of spiritual, psychological, and philosophical traditions, and go straight for what really counts: *growing those neural bundles to the tipping point.*

While some of the techniques taught by those traditions are helpful in growing those neurons, others are not. Using MRIs and EEGs, we can now hone in on which practices truly spark synaptic growth and those that are just window dressing. The four techniques selected from the *Yoga Sutras of Patanjali* and used by the Alzheimer's patients we met in Chapter 2 are an example of effective methods. So are the parallel techniques used in EcoMeditation. They bulk up the neural circuits of SQ quickly and take you directly to the tipping point. Neuroscience shows us the way.

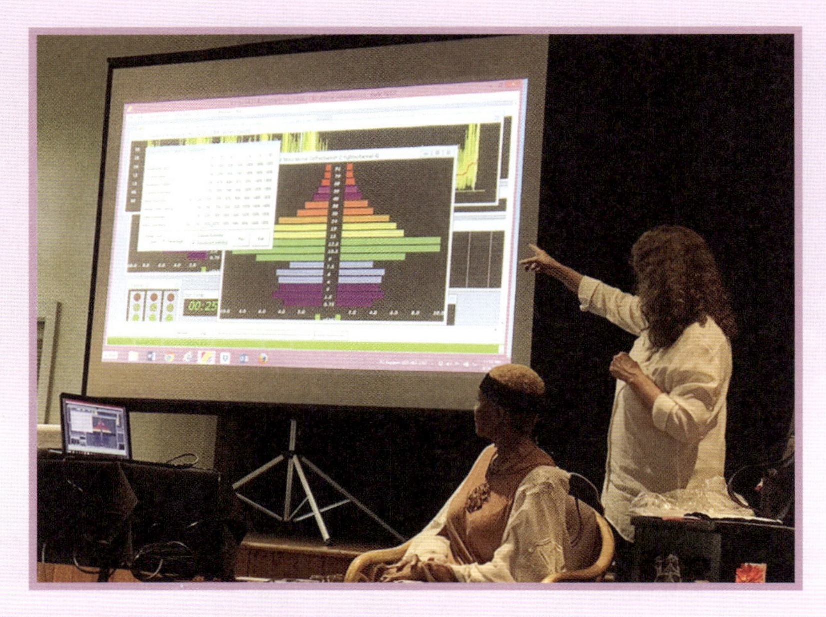

3.6. Brain scientists can now determine which practices are most effective at building the hardware of SQ in your brain.

Spiritual Intelligence as a Personality Trait

People who have the trait of SQ don't just experience it as a temporary state of mind. It's woven into the fabric of their characters. Think about a person from history who you admire for the good they did in the world. They could be depended on to inspire and uplift those around them. They were change agents. They exemplified SQ under all circumstances, favorable or adverse. That's what SQ looks like as a dependable character trait.

Look at the list you made in Chapter 1, as you thought about people who embody SQ. If you reflect deeply, you'll realize that they had the trait of SQ; it was not just a passing state. They might have had ups and downs in their level of spiritual connection with that source greater than themselves, but they tended to maintain that connection reliably. That's what the trait of SQ looks like. After they reached the tipping point, SQ became hardwired into their brains.

It's analogous to the straw that broke the camel's back. The porter adds thousands of straws to the beast's burden and the camel is able to carry them all. Then just a single new straw is added and "the last straw" is too much weight for the camel to sustain. Suddenly and surprisingly, the tipping point is reached. Once it's attained, everything is different.

Once we pass the tipping point and SQ becomes embedded in our neural wiring, it shapes everything we do and experience. There are stories of people who faced the harshest of extreme circumstances but did not lose their compassion, emotion regulation, and other characteristics of SQ.

3.7. People who demonstrate the trait of SQ in the face of adversity inspire others.

One of the most famous of these is Victor Frankl, the Austrian psychiatrist who was imprisoned at Auschwitz and later told his story in his book *Man's Search for Meaning.*[15] But WWII fostered many other such heroes as well. One of my favorite stories is that of Etty Hillesum, a Dutch woman who maintained SQ in the face of the most barbarous inhumanity imaginable.

THE MYSTIC OF THE HOLOCAUST

Etty Hillesum (1914–1943) is called the Mystic of the Holocaust.[16]

She was a young woman who resided in Amsterdam during the Nazi occupation and perished as one of the millions of Holocaust victims. Throughout her brief but extraordinary life, Etty came to perceive God within the depths of her soul and in others.

3.8. Etty Hillesum.

Etty was born into a Jewish family, many of whom struggled with depression and illness. She earned a master's degree in law from the University of Amsterdam. Etty became a patient of Julius Spier, a German psychoanalyst who was also her lover for a time; he was the spiritual mentor who guided her toward inner transformation.

As part of her therapy, Etty began keeping a journal, documenting her practice of "soul-listening," and providing a detailed account of her journey to SQ. She began to see life from a new perspective and dedicated herself to serving Jewish refugees in a Nazi transit camp.

Etty meticulously recorded her daily experiences and internal reflections in her diary for the last two years of her life. From the imposition of the yellow star on Dutch Jews to her deportation to Poland, Etty dedicated herself to an ambitious mission. Facing imminent death, she aimed to bear witness to the unbreakable power of love and to reconcile her sensitivity to human suffering with her appreciation for the beauty and meaning of life.

Etty remained deeply connected with the Jewish community. However, her reflections were enriched by a diverse range of sources, including Rilke, the Bible, St. Augustine, and Dostoevsky. When a friend indignantly remarked that her views on loving one's enemies resembled Christian beliefs, she replied, "Yes, Christianity, why ever not?"

Despite this, Etty had little interest in organized religion. In an era when "the whole world is becoming a giant concentration camp," she believed it was crucial to cling to what is enduring—the encounter with the infinite within one's own life and that of others.

Ultimately, Etty and her family were caught up in the horrors of the Holocaust. They were transported to Poland by a death train and Etty died in Auschwitz on November 30, 1943, at the age of 29.

Friends of Etty preserved her journals for decades, struggling to get them published. Finally, in 1981, they succeeded and the journals were published in Dutch. Two years later, they appeared in English as *An Interrupted Life*.[17] This account became a powerful testament to spirituality and resistance against persecution and hatred, showcasing Etty's struggles, prayers, growing Spiritual Intelligence, and acceptance of death.

It is unlikely that you and I will ever be tested as severely as Etty Hillesum or Victor Frankl. But we all face adversity in our lives. I've lived through many difficult times.

These include the death of my second son, Montague, a story which I tell in my book *Facing Death, Finding Love.*[18] In 2017, my home and office were swept away in minutes during a California wildfire and my wife, Christine, and I barely escaped through the flames. That story is told in the opening chapter of my book *Bliss Brain.*[19]

Twice in my life—once in my forties and once in my sixties—I lost all my savings and began rebuilding my fortunes from scratch. I was in several tumultuous romantic relationships before meeting Christine. My mother died young from cancer and my father spent 15 years gradually disappearing into the fog of Alzheimer's disease before his body gave up the ghost.

You probably have similar stories. So even "ordinary" people like you and me have a fair amount of challenge in our lives.

But when you've changed happiness from a state to a trait and used SQ to build a resilient brain, you bounce back. You find yourself happy no matter what. You have an inner equanimity that cannot be shaken by outer circumstances, no matter how dire. Stories like that of Etty Hillesum remind us that once we've passed the tipping point, we are the masters of our inner world, even though our outer world may be in chaos.

From Seeker to Finder

My dear friend Jeffery Martin has studied people who awaken into the state of oneness. He has evaluated the experience of over 2,000 people who have had enlightenment experiences and identified the most common characteristics of this type of experience. He summarizes these in his book *The Finders.*[20] The term "Finder" refers to crossing the boundary from being a "Seeker" of truth to being a Finder.

Many of us began our journey as "Seekers" decades ago. We may have been Seekers of meaning in our lives, aspiring to understand the answers to life's greatest mysteries, to know the nature of the universe, to find God, to identify an authentic spiritual master, to discover the secret of happiness, to unlock our full human potential, to experience elevated spiritual states, or all of the above.

For 20 years, Jeffery was obsessed with people who'd come to the end of the quest. They'd *made the transition from Seeker to Finder.* They'd reached the tipping point.

After collecting hundreds of interviews with Finders, some of them days in length, he began to notice patterns.[21] While they differed in some important ways, Finders also had distinct commonalities. Among these are:

- A sense that everything is fundamentally okay
- A reduction in self-absorbed mind chatter
- Centering in the present moment
- Little interest in the stale old stories told by others and their own minds
- A shift in negative emotions; while they still arise, they are transient
- A sense of connection between inner experience and the greater reality
- Reduced attachment to previous goals and outcomes

Doesn't that sound like a wonderful way to live your life? Imagine if—despite all the turmoil in the world around you—your SQ was so strong that you felt a basic sense of "okayness" about yourself and everything else.

Imagine your habits of worry and stress being quieted by this sense of okayness. Jeffery calls it "fundamental wellbeing." That's a sense of wellbeing that's fundamental to your view of yourself, of others, of the world, and of the universe. It's a rock on which you can build positive emotion.

Characteristic number two is related to fundamental wellbeing and it's a quieting of the incessant chatterbox inside our heads. This is the voice that has something to say about everything going on. Another of my friends, Steven Kotler, calls it "your inner Woody Allen."[22] From the moment we wake up, this voice is commenting, judging, criticizing, evaluating, finding fault, complaining, and looking for the downside.

Jeffery calls it the "Narrative Self." It maintains a running commentary on everything we do, others do, the world does. He defines the Narrative Self as "the self-referential, story-based form of self that houses the collective past and forms the basis for identity creation and maintenance."[23]

In Finders, the Narrative Self quiets down. This gives them a degree of inner peace and relief from the flow of mental chatter and mind wandering that constantly distracts us from the practices that build SQ.

This opens the door to characteristic three, being in the present moment. The Narrative Self is rarely there. It's always in either past or future—Caveman Brain obsessing about the hurts and wounds of the past or catastrophizing about imagined calamities in the future. When this activity dials down, we enter the present moment. That's the only place we can find love, joy, and peace.

When we're in the present, old stories and self-concepts become less sticky and we open to a view of ourselves and others that is fresh, vibrant, and full of potential. We're less interested in where we've been than where we're going. Our old stories begin to sound boring.

While Finders still experience negative emotions, they dissipate faster. They also have a sense of connection between what they're experiencing in their inner lives and a greater reality. That's the essence of SQ.

The final characteristic is less attachment to rigid outcomes and fixed goals and a greater sense of dancing with the flow of life in each moment. We'll look deeply into the inner experience of Finders in Chapter 4 and you'll discover that you've likely already experienced profound mystical states, perhaps without even knowing it.

PRACTICING THE PRESENCE

In 1614, Nicholas Herman was born into a peasant family in Lorraine, France. In his early teenage years, he followed the path of many young men seeking to escape hunger and poverty and joined the army. King Louis XIV was on the throne and the Thirty Years War (1618–1648) meant soldiers were in high demand.

But Herman was soon captured by the Germans. Abused and threatened with hanging, his faith was so strong that he "viewed death with indifference," according to his biographer Father Joseph de Beaufort. After his release, he returned to the army but was injured in a battle with Sweden, leaving him permanently disabled. Beaufort reported that Herman "often relived the perils of military service." Today that rumination would likely result in a diagnosis of PTSD.

After Herman's discharge from the army, he took a job as a footman to William de Fuibert, treasurer to the king of France. But his clumsiness soon resulted in his discharge from his new profession.

On a winter's day when he was 18 years old, Herman was gazing at the barren stalks of a frozen tree. Suddenly, he saw the tree as it would be in a few weeks, with leaves and flowers erupting in a wild frenzy of exuberant rebirth.

Beaufort writes, "Considering that within a little time, the leaves would be renewed, and after that the flowers and fruit appear, he received a high view of the providence and power of God, which has never since been effaced from his soul."

In that moment, Herman flipped from being a Seeker to a Finder. He decided to dedicate his life to prayer and devotion. Herman joined the Carmelite monastery in Paris in 1642 at the age of 26. As his patron, he chose third-century martyr St. Lawrence (225–258), becoming Brother Lawrence. He added the words "of the Resurrection" to his official name to honor his tipping point.

Brother Lawrence's subsequent career was neither heroic nor prestigious. He is remembered today not for brilliant theological insights, impassioned preaching, or monumental social contributions, but for his simple spiritual practice. Uninspired by the rigid rules and elaborate devotions of the Carmelites, he focused simply on *constant awareness of God's presence.*

Beaufort compiled Lawrence's letters. In one, Lawrence writes, "I make it my business only to persevere in His holy presence, wherein I keep myself by a simple attention, and a general fond regard to God, which I may call an actual presence of God; or, to speak better, an habitual, silent, and secret conversation of the soul with God."

His fellow monks didn't regard Lawrence as having much in the way of intellect or useful skills. So they stuck him in the place he liked the least, the monastery kitchen. There he spent the next 30 years, cooking and scrubbing the grime off the pots.

3.9. Brother Lawrence.

Yet Lawrence decided to embrace his menial work fully. He turned it into an act of devotion, declaring, "Nor is it needful that we should have great things to do.... We can do little things for God. I turn the cake that is frying on the pan for the love of Him; and that done, if there is nothing else to call me, I prostrate myself in worship before Him Who has given me grace to work. Afterwards I rise happier than a king."

Lawrence also wrote, "I have no will but that of God, which I endeavor to accomplish in all things, and to which I am so resigned, that I would not take up a straw from the ground against His order, or from any other motive but purely that of love to Him."

He told a woman who came to visit the monastery, "We do not have to be in church to be with God. We can make of our hearts an oratory where we can withdraw from time to time to converse with him, gently, humbly, and lovingly. Everyone is capable of these familiar conversations with God, some more, some less..."

Beaufort's compilation of letters was eventually published as *The Practice of the Presence of God*. It has sold some 22 million copies, making it one of the best-selling books of all time. You can find it in the Extended Play section at the end of this chapter.

Lawrence ended his last letter with the words, "I hope for the merciful grace of seeing God in a few days." Lucid up to the last moments, Brother Lawrence died at the age of 77.

All these centuries later, Lawrence's humble reminder endures: "Lift up your heart to Him, the least little thought of Him will be acceptable. You need not cry very loud; He is nearer to us than we are aware of."

Lawrence's tipping point occurred when he looked at a tree and saw how it would look in spring. Before that, he'd had spiritual experiences, building up his neurological "muscles" to the point where they could support a breakthrough in consciousness.

After the tipping point, Lawrence was able to take even adversities like being banished to the monastery kitchen and turn them to his advantage. As he flipped cakes and scrubbed pots, he communed in "secret conversations" with God. This resulted in such bliss that Beaufort and other monks were inspired by Lawrence's presence.

Jeffery's study of thousands of Finders shows that awakening brings with it an "unimaginable level of contentment and wellbeing (even if you already think of yourself as very happy)."[24] Like Brother Lawrence, once you pass the tipping point, everything is different.

Finder Minnie LaBelle puts it this way: "Awakening profoundly changed my outlook on everything—overnight. Fears, anger, disappointments, and misunderstandings that had driven my behavior for decades all faded rapidly.

"Nothing changed, yet everything changed. I still looked like the same person on the outside, but I saw the whole world without judgment or stories. I remained perfectly calm whatever transpired. That was 30 years ago, and while occasionally I get a bit triggered by an upsetting event, it's like a mist that just blows away."

When Minnie awakened, her whole life changed literally overnight. But she had been a Seeker for decades before she became a Finder. In those long years, she built up the synapses in her Enlightenment Network that eventually carried her past the tipping point.

The Long Path to Enlightenment

Spiritually intelligent people have been striving to achieve elevated states for millennia. And throughout history, the great spiritual traditions laid out a clear path to enlightenment: You abandoned ordinary life, joined a religious order, took vows like poverty, chastity, and obedience, pursued mystical experiences, and spent decades in spiritual practices. That's the Long Path.

The 10,000 hour rule is the concept that to master a skill, you need to spend an extended period of time practicing it.[25] Surgeons, dancers, pilots, teachers, gardeners, musicians, accountants, athletes, and other professionals all get better with practice, taking the Long Path to expertise.

It's not a matter of simply repeating an activity that builds expertise. That 10,000 hours must be spent in "highly structured activity, the explicit goal of which is to improve performance."[26] Assuming 40 hours a week, the Long Path translates into over five years spent in carefully structured performance-oriented activity.

That's why researchers pay such close attention to meditators like Mingyur Rinpoche whom we met in Chapter 2. They've spent over 10,000 hours developing SQ through the highly structured approach derived from thousands of years of tradition in Tibetan Buddhism.

The Long Path is one of self-improvement. It involves steps, techniques, initiations, skills, and self-purification. It is a gradual and incremental process.

The Long Path uses practices to advance the Seeker's spiritual development. These take the form of physical exercise such as yoga or Qigong; dietary restrictions like vegetarianism, fasting, and abstinence; mental training in various forms of mindfulness; study of the Scriptures; obedience to a spiritual teacher; moral awareness; vows; and other forms of discipline that develop SQ.

For laypeople who are not monks or nuns, the Long Path has many of the same elements, but structured around ordinary life. Devotees may spend one day out of seven at the temple while taking care of family, self, and business the other six. They may read sacred books in their leisure time. They may sincerely seek to live by a moral code. They may modify their eating and drinking habits to avoid excess. They may donate money and volunteer time in the service of good causes. Such Seekers do their best to develop SQ among the noise of everyday life.

They won't get even close to the 10,000 hours of expert meditators in their entire lifetimes. However, research shows us that the breakthrough to SQ is still within their reach.

From Religion to Psychology

In the 19th century, science—and particularly psychology—emerged as an alternative to the traditional religions. People began to see a path to transformation through reason and the mind. It was no longer necessary to trust authoritarian religious figures or follow strict archaic rules.

The behavioral psychologists of the early 20th century showed that most behavior is conditioned. The most famous of these was Ivan Pavlov. He conditioned dogs by ringing a bell when they were being fed. They learned to associate the sound of the bell with the appearance of food. After conditioning was complete, they would salivate at the sound of the bell even when no food was present.

In humans, it is equally obvious that much of our behavior is conditioned. This includes suffering. Think about friends or family members who suffer from self-inflicted mental wounds. They have thoughts, beliefs, habits, and behaviors that lead to unhappiness every single time. Yet they keep repeating them.

So the next obvious question psychologists asked was whether they could "countercondition" the negativity bias of Caveman Brain and liberate people from stress.

The term "stress" was coined by Hungarian doctor Hans Selye. As he worked in hospital wards, he noticed a collection of symptoms common to patients regardless of their diagnosis. These included increased heart rate and blood pressure, muscle tension or pain, disruptions in sleep patterns, changes in appetite, difficulty concentrating or making decisions, and feelings of anxiety, irritability, or depression.[27]

In the early 1970s, Herb Benson of Harvard University developed a method called the Relaxation Response. It was a program explicitly designed to countercondition the stress response that Selye had described.

HERB BENSON AND THE RELAXATION RESPONSE

3.10. Herbert Benson, MD.

Herbert Benson (1935-2022) was a pioneering doctor whose research connected mind, body and SQ. His career was marked by his innovative efforts to bridge the gap between Western medicine and traditional holistic healing practices based in breath work, time in nature, music, and art.

He published his seminal work on the Relaxation Response in 1975.[28] The Relaxation Response is a stress-reduction method designed to counteract the physiological effects of stress and the fight-or-flight response identified by Hans Selye.

Benson's approach is grounded in the principle that through certain practices, many of them drawn from ancient traditions, one can invoke the body's own natural relaxation mechanisms, leading to a state of rest and balance. The core components of the Relaxation Response include:

- **A Quiet Environment:** A setting free from distractions, where one can relax without external interruptions.
- **A Mental Device:** A focus of attention such as a word, sound, phrase, mantra, or repetitive prayer. This element helps to anchor the mind, allowing it to move away from distracting thoughts.
- **A Passive Attitude:** The release of worry about how well you are performing the technique, or anything else. Instead, you allow thoughts to come and go without judgment or engagement.

- **A Comfortable Position:** Ensuring physical comfort to minimize distractions and facilitate relaxation.

The combination of these elements significantly reduces both physical and psychological stress. This leads to decreased heart rate, blood pressure, and muscle tension, among other benefits. The Relaxation Response has been widely adopted and adapted across various contexts, including stress management, therapy, and wellness programs, highlighting its effectiveness and the pioneering nature of Benson's work in the field of stress reduction and integrative health.

If you look closely at those four components of the Relaxation Response, you'll see they bear a close resemblance to the techniques from the *Yoga Sutras* of Patanjali practiced by the Alzheimer's patients in Chapter 2, as well as the components of EcoMeditation. They straddle modern psychology and ancient spiritual practice.

As psychology has offered increasing promise of an escape from our suffering, formal religious observance has been declining. In the 19th century, atheists were rare and virtually all members of societies engaged in religious practice, whether devout or perfunctory. The percentage of such people declined steadily in the 20th and 21st centuries. Today, in Western Europe, only 22% of people attend church services monthly or more.[29] The figure for the US is 30%.[30]

Today, we're much more likely to look to science for answers to the questions of meaning, happiness, and fulfillment. Psychology has taken much of the space formerly occupied by religion. Body-based methods like Herb Benson's Relaxation Response have brought relief from suffering to millions of people who might never cross the threshold of a church or walk the Long Path.

Neuroscience Trumps Psychology

Psychology is the study of the mind. Neuroscience goes deeper, to study the brain, the substrate that anchors the mind in the body.

Neuroscience is in the process of producing a quantum leap in our understanding of personal development. The traditional Long Path approach took at least 10,000 hours to show results. Despite the ascendance of psychology, the World Health Organization finds that anxiety and depression are rising globally, so we're losing ground in our quest for happiness.[31]

But bypassing spiritual and psychological methods and explanations, neuroscience shows us we can change the brain directly. It doesn't rely on traditional spiritual formulas or psychological theory. Researchers can hook up a person to an EEG or MRI and find out the precise effects a practice has on the brain. It shows us how we can develop the Enlightenment Network, reach the tipping point, and go from being a Seeker to a Finder.

In the coming decades, the techniques we use to guide our life journeys, including the development of SQ, will be drawn from neuroscience rather than psychology or spirituality. This is how we get to the tipping point as quickly and efficiently as possible.

The Short Path to Enlightenment

As a teenager, I went to live at a spiritual community. The other members and I were dedicated to the Long Path. We grew our own food, meditated, studied the great spiritual traditions, followed the instructions of the teachers, experimented with a vegetarian diet, tried to be compassionate to each other, and admired those who had become enlightened.

But on the Long Path, I heard intriguing rumors about a Short Path. This was interesting because I was far from happy living in the community. And it was hard to avoid the realization that hardly any of the people around me were happy either, even though we were doing all the "right" things. Our austerity was not producing relief from suffering.

Not only did a life of discipline and misery loom ahead of me, but according to the Eastern scriptures, one lifetime was rarely enough. It might take hundreds of lifetimes over thousands of years to "work off your karma" and reach the tipping point. A Short Path sounded intriguing.

I found it in the work of Paul Brunton. This remarkable mystic from a century earlier collected the Short Path teachings of many of the great spiritual masters of his time and presented these to Westerners in a form they could understand and practice.[32]

A SEARCH IN SECRET INDIA

Paul Brunton was born in England in 1898. He usually referred to himself as "PB." After World War I, he began traveling in search of the answers to life's great questions. He was already a seasoned meditator and he'd had extraordinary mystical experiences, though he did not yet understand them.

3.11. Paul Brunton.

PB's quest took him first to India, then to Egypt, and eventually all over the world. At different times, he lived in Japan, China, Greece, California, Ohio, New York, Bolivia, Mexico, Switzerland, Italy, Spain, Austria, New Zealand, and Australia. He sought out the great spiritual masters of every religion.

Brunton's first major book, *A Search in Secret India,* was published in 1934.[33] Eventually, he wrote many others, including *The Wisdom of the Overself, The Hidden Teaching Beyond Yoga, The Quest of the Overself, The Secret Path,* and *The Inner Reality.*

During the 1930s, his books sold millions of copies, and they have continued to provide guidance for Seekers ever since. They have been translated into over 20 languages. After his death in 1981, the Paul Brunton Philosophic Foundation published, in 16 volumes, *The Notebooks of Paul Brunton*.

PB kept pen and paper with him at all times. After he had spent a day or a year with a saint or mystic, he would write down the teachings he had received as well as his own inspired reflections. He understood the commonality of the mystical experience that transcends all religions and, with his vast understanding of SQ, he connected the dots between them.

Due to his reputation as a philosopher and sage, Brunton gained access to teachers and teachings that had previously been unavailable except to initiates.

He became close friends with Ramana Maharshi, a great Indian teacher of "nonduality" or oneness with the divine.

Brunton also inspired many of the great teachers of our time. Human potential pioneer Jean Houston said PB was "a great original and got to a place of personal evolution that illuminates the pathways of a future humanity." Professor Kenneth Ring described PB's writing as "the acme of wisdom on the nature of human spirituality."

Nondualism teacher Adyashanti observed that PB's book *The Short Path to Enlightenment* contains "profound teachings of immediate spiritual awakening that have the power to short circuit the seeker in us and reveal the true nature of reality here and now." One of my favorite contemporary mystics, Gangaji, said this book is "alive with supreme knowledge."[34]

When Christopher Reeve, the actor who played Superman, was asked which book he would choose if stranded on a desert island, he chose Brunton's *The Inner Reality*.

I found Paul Brunton's books inspiring when I was a teenage Seeker, but I had a hard time understanding his work. But when I circled back to Brunton in my sixties, I found him one of the most articulate proponents of the Short Path.

Recognizing the Self Within

The essence of PB's Short Path teaching is to release focus on the ego purification inherent in the Long Path. Instead, you turn within. You direct your attention to your Higher Power, which PB calls the "Overself," and to infinite universal consciousness itself.[35]

Instead of struggling to be better and escape the grip of suffering, as I was doing as a teenager in the spiritual community, PB has us focus on the present moment. Mahayana Buddhism reminds us that our core nature is the Buddha, the enlightened one. Ramana Maharshi called it the Self with a capital "S" and we experience it only in the present moment.

Enlightenment isn't a state you attain, the end stage of a long spiritual journey. *It's the true nature of every single human being.* Awakening isn't becoming aware of a transcendent new reality. It's simply waking up to what you already are, your true Self, the Buddha that dwells within.

PB taught that the Long Path of the Seeker can actually perpetuate the false self, the ego. When you let go and surrender to the infinite, you take the Short Path, opening yourself to grace. That grace can awaken us to the Self that's been inside us all along and take us to the enlightened state. We go from Seeker to Finder. Grace can even descend mysteriously at any moment of our walk along the path. There are rare stories of people who, without any spiritual training or intention, suddenly awakened, carried past the tipping point by grace alone.

Though PB taught the Short Path, his work emphasizes the value of the Long Path. We need to discipline our urges, refine our character, and prepare our minds for the Short Path. The Long Path accomplishes this. Even after the Short Path has taken us to Self-realization, we continue the practices of the Long Path. We continue to take care of ourselves and the world around us and purify our character. The Short Path and the Long Path are not alternatives. They're complementary.

Walking the Short Path

In his book *The Short Path to Enlightenment,* PB uses the term "milestones" to describe stages of spiritual development. I got excited about these a few years back because I realized that they paralleled phases of brain change in SQ, as we'll see in Chapter 6. Brunton also created "exercises" based on the teachings of the masters. Practicing these exercises furthers our spiritual journey.[36]

My research using EEGs and MRIs had shown me that certain of these exercises reflected the precise methods that produced rapid brain change. The masters had discovered them through thousands of years of trial and error; they were now being confirmed by scientists examining the inner workings of the brain.[37-38]

In 2021, I went on a long retreat and began turning PB's milestones and exercises into a series of courses that anyone could take. I cataloged all the milestones and found that there were 36 of them. I arranged the exercises to take Seekers from milestone to milestone. You'll find links to these courses at the end of this chapter. The most comprehensive of these is called "The Short Path to Oneness."

Thousands of people have now taken these courses. Research shows that they produce leaps in happiness and ecstatic self-transcendent states. They dial down anxiety and depression and make people much more productive in their everyday lives.[39]

Here's a story by one of the participants in the Short Path to Oneness. Greg persevered through easy times and rough patches, and eventually experienced big shifts.

TRANSCENDING THE MASKS WE'VE WORN

By Greg Berg

Is it really possible to find and then maintain a greater baseline of joy and peace and wisdom amidst the inevitable ups and downs of life? Are there easy practices that can rewire our brain, accelerate our growth and healing, and help us tap into higher states of consciousness and unity more often?

3.12. Greg Berg.

The answers to those questions form both the premise and the journey for participants in Dawson Church's "The Short Path to Oneness" course. After years of trying various healing modalities and practices that offered only glimpses of something greater and having already been familiar with Dawson's work, I was willing to jump down the rabbit hole and find out.

Nearly two years after beginning the Short Path process, my personal answer to each question is yes... with an asterisk.

*It takes commitment and patience and a willingness to look in the mirror in new and deeper ways.

But if we say "yes" to all of that, we can indeed begin to see and transcend the plays we've been participating in, the masks we've worn, and the patterns we repeat. We can move beyond our habitual, limited ways of being into something greater—what the great sages have called the infinite. With the Short Path as our guide.

I can't claim to be living a life without any challenges. I still feel the bumps in the road from time to time. But as someone who's battled depression and struggled with how I moved through the world at times, I have sensed a shift in myself after doing this work.

I can see temporary states and my "local self" for what they are; I can more easily sit with and reframe discomfort or challenges; and most importantly, I can access a Universal intelligence (my "higher self") that allows me to be more present to the now and helps guide my way forward.

So thank you, Dawson, and all of your contributors and collaborators and guides for sharing this wisdom with the world. The Short Path to Oneness is truly a gift.

As Greg's story shows, the journey to SQ isn't a weekend trip. But using neuroscience-based techniques as well as ancient Short Path practices, we make steady progress. With consistency, we can get to the tipping point in a year or two. It doesn't have to take the "thousand lifetimes" referred to in some of the Eastern scriptures.

Chop Wood, Carry Water

The tricky thing about the tipping point is that you don't know when you're almost there. You might have built up 99.99% of the neurons required for change, yet you are still just experiencing temporary states.

That's why seers like Paul Brunton and Ramana Maharshi recommend you stick faithfully with your Long Path practices. They pave the way for the Short Path.

Even once you're a Finder, living from the Self within, with your brain's Enlightenment Network firing every day, you maintain your spiritual practice. You meditate, live mindfully, act ethically, and demonstrate SQ with each thought, word, and deed.

3.13. Before enlightenment? Chop wood, carry water. After enlightenment? Chop wood, carry water.

Neuroscience has now taken us far beyond the guidelines of traditional spirituality and psychology. It's shown us exactly which practices are effective at growing SQ, based on MRI and EEG research. We are in a new era in which we can cultivate a spiritually intelligent brain in a matter of weeks, not 10,000 hours. It makes the elevated transcendent states of SQ available to everyone, not just long-practicing adepts. It turns these states into durable traits.

SQ leads us to elevated experiences of bliss and oneness. But these mystical states are the app. The hardware that runs them is the Enlightenment Network. When the circuits in this network get big enough, we reach a tipping point at which we have enough hardware in our brains to run the full software package of SQ. We then graduate from a Seeker to a Finder. In the next chapter, we'll examine some of the extraordinary experiences this brings.

Deepening Practices

The Deepening Practices for this chapter include:

- **Journaling Exercise 1:** Look back at your own life and identify key moments of spiritual growth. Record the title and year of each of these in your journal. This is the outline of your "spiritual autobiography."
- **Journaling Exercise 2:** What tipping points have you reached in your life? These could have to do with career, relationships, money, health, or spirituality. When, after practice, did you suddenly jump to a new state? Record at least three in your journal.
- **Inspirational Reading:** Chapters 1 and 2 of Dawson's Short Path to Oneness course in the Extended Play section
- **Dawson Guided EcoMeditation:** Observing My Thoughts

Get the meditation track at: SpiritualIntelligence.info/3.

Extended Play Resources

The Extended Play version of this chapter includes:

- Transcript of a talk by PB describing the Short and Long Paths
- Chapters 1 and 2 of Dawson's Short Path to Oneness course
- Podcast: The Awakened Brain: Lisa Miller and Dawson Church in Conversation
- Guided version of the Relaxation Response from Mount Sinai Health System
- Register to find out more about the 21 Day Walk With Your Higher Power virtual course and the Short Path to Oneness yearlong Master Mind at TheShortPath.com.
- Download Brother Lawrence's *Practice of the Presence of God.*

Get the extended play resources at: SpiritualIntelligence.info/3.

Updates to This Chapter

As new studies are published, this chapter is regularly updated. Get the most recent version at: SpiritualIntelligence.info/3.

Chapter 4

The Awakened Experience

"Most men live lives of quiet desperation." So said Henry David Thoreau two hundred years ago. Thoreau's book *On Walden Pond* presented an idealized picture of a tranquil life. It inspired millions of people with its message of a simple life stripped of unnecessary materialism. It celebrated self-reliance, simplicity, beauty, and nature.

But the dark side of Thoreau's experience is known by few people. He went through several periods of "quiet desperation" himself. He kept a daily journal for over 20 years and these show he suffered bouts of deep depression. On several occasions, he came to the very brink of committing suicide.[1]

Then there are the masters, living each moment in peak self-transcendence. They're at the other end of the scale. Research shows they're happy. More than happy, they're hitting ecstatic states so elevated they burst through the envelope of current scientific knowledge. In this chapter, we'll explore the ecstasy they inhabit: the awakened experience.

What is that experience like? In Jeffery Martin's study of people who go from being Seekers to Finders, who pass the tipping point and discover the Holy Grail of enlightenment, he finds a consistent pattern of "fundamental wellbeing." That's the unshakeable bedrock conviction that, no matter what happens inside of me or outside of me, everything's basically okay.[2]

In my conversations with Jeffery, for the term "fundamental wellbeing," we've played around with alternatives like "extraordinary happiness." The research reviewed in Chapter 3 showed that when feeling compassion, the brain waves of ecstasy jump as much as 25-fold. In this state, people aren't 25% happier. They're *25 times* happier.[3]

Our modern languages like English don't even have words to describe a condition this happy. We have to resort to Sanskrit, the language of Patanjali and other ancient masters, to find words that designate the moods of the masters.

Just the way Eskimos had 50 different words for "snow" or native Hawaiians had over 200 words for "rain," the Sanskrit language has many different words for what we term "happiness." Like *vishoka,* which means "sorrowless joy." And *ananda,* which means "ecstatic bliss."

In Sanskrit, our conventional conditional happiness is at the bottom of the scale, encapsulated in the word *preya*. When you go far beyond preya and reach a sustained state of happiness, that's termed *sukha*. Vishoka is experienced only in rarified states far above the experience of non-meditators, and ananda by masters.

In this chapter, we'll go inside the experience of those who've soared to those lofty heights of feeling. Who've gone from being Seekers to being Finders. Who've abandoned Caveman Brain for the ecstasy of Bliss Brain. Who've tamed the Default Mode Network and activated the Enlightenment Network. Who've made it all the way to vishoka and ananda.

For decades, researchers have been fascinated by this lucky few. They've studied their common characteristics and we now have a full understanding of the awakened experience.[4] Looking into their minds and hearts will inspire you with the possibilities of sharing those states.

The following story is a great example of someone who had a transcendental experience and it came unexpectedly out of the depths of despair.

THE DAY EVERYTHING CHANGED

By Dawn Woodwardson

The walls of Awa'awa'puhi Valley rolled out in front of me. The dark green hues of trees, the light greens of undergrowth, the ochre shades of clay, the dark browns of bedrock. The ripples of the mountain, pleated like the folds of a tablecloth.

At my feet, tiny purple flowers grew from thick-leaved succulents. The limbs of long-dead trees surged up between clumps of living foliage, twisted into grotesque shapes.

The sun played across the scene. Behind flower petals, it made them glow from within. It lit the tops of the valley walls with brilliance, plunging the valleys into deep shadow. Millions of years of erosion had uncovered the horizontal strata of the rocks—sedimen-

4.1. Awa'awa'puhi Valley.

tary, igneous, metamorphic—laying the evolution of Mother Nature bare for all to see.

I lost myself in the scene. Time and place fell away. The beauty and perfection of the moment captivated me utterly. A huge peace filled my heart. The tragedies of my life and the worries of my mind disappeared, swallowed by the moment.

Tears began to roll down my cheeks. Tears of gratitude, humility, awe. The hugeness of the perfection of the universe overpowered my puny sense of human struggle.

My body sank slowly to the ground. I felt all my weight settle into the ground, as though Mother Earth herself was cradling my body. I stretched out my arms and placed the palms of my hands in the soil and scratchy undergrowth. The presence of life itself surged up through my arms, like an electric fluid filling up my body.

The plants around me seemed to glow with an aura of light. The dirt itself felt alive, pulsing with the energy of the planet. I watched an ant, crawling up the stem of a flower. Each of its six legs moved with exquisite grace. The light was in the ant, the light was in the plants, the light was in the soil, the light was in my fingers. Rather than solid matter, everything seemed to be shimmering patterns of energy.

Then I thought about all the tragedies of my life, and uncontrollable sobs began to rack my body. I stretched out full length on the grass and cried like my heart was breaking. Something in the core of my being cracked wide open. The old me just shattered. All of the fear and pain that I'd been holding gushed out through my eyes, and the ground seemed to soak it in. I must have been sobbing on the ground for 15, 20 minutes. It seemed like the tears would never stop.

I'd come here to the Hawaiian island of Kauai to escape that past. My husband Anton and I had just gone through a bitter divorce after 15 miserable years of marriage. During the long divorce proceedings, we fought about everything, just like we had during our marriage. Our real estate business, our 12-year-old son Jackson, our house, our cars, our possessions.

I was exhausted, mentally and physically. The business had been so damaged by our marital disputes that it was almost destroyed. Lawyer's fees had eaten up all our savings plus the equity in the house Anton and I owned. We were both broke, so I was starting from scratch. Jackson was acting out in school and at home. From nowhere, horrendous migraine headaches would hit me several times a month. The pain was so severe that I couldn't stand any light. My eyes filled with a red mist and I couldn't drive. I gained weight and my doctor told me I'm prediabetic. I'm pushing 40 and the strain was showing in my face and body.

So here I was, on this February day, hiking the crest of Waimea, the "Grand Canyon of the Pacific." I'd hiked an hour out from the Koke'e Visitor Center when I turned a bend in the trail and saw Awa'awa'puhi Valley.

And here I was, stretched out on the ground, sobbing uncontrollably. I thought the tears would never stop.

But they did, and a huge peace came over me. I was empty.

I felt connected to everything. I stood up and looked around me, at the grass, at the flowers, at the impossible jagged peaks.

As the beauty of the scene flooded through my body, as I saw the whole universe as light, I knew I was seeing the "real" reality. All my problems seemed like meaningless illusions.

I knew in my bones that everything was OK, always had been, always would be. That I was part of what I was seeing, a human speck in a beautiful universe.

I began to shake with ecstasy. I laughed through my tears. Laughing while crying. Crying while laughing.

My heart swelled to where it felt like it was bursting through my ribs. Love and gratitude poured into me from the universe, far beyond my ability to receive. I thought of everyone in life, those I'd loved and those I'd fought with. Those that had caused me pain and those that had brought me joy. All the confusion and struggle of my relationships. All my doubt about the meaning of life—my weird and unpredictable life in particular.

As I thought again about my past and all the horribleness of the last few years, it seemed like a dream. I didn't feel any attachment. I was at peace with it all, just the way it was. I knew it had meaning, whether or not I could understand it. I realized it had taken all the tragedies of my past to strip away the things I believed were important, freeing me up to see the only thing that is truly important, the love in the universe and the beauty of the present.

This wasn't anything like the God I'd learned about in Sunday School. It was the energy inherent in the universe. And it wasn't a mysterious energy, it was more solid than reality. More real than the matter of which my body and the earth were composed. And it was love, pure love. The love filled my body and my past.

I sensed that I would also be OK in the years to come. All my fears for the future—of heartache, conflict, loneliness, poverty, meaninglessness, boredom, frustration, aging, disease, and death—fell away. Not only did my past make sense, like the pieces of a jigsaw falling into place. My future made sense as well, even though I had no idea what it might hold.

I have no idea how long I stood there staring at the canyon. Eventually, the experience felt complete and I got up slowly and hiked back to my car. But everything was different. I felt a calm that stayed with me for weeks on end. I felt strong enough to handle the challenges that would face me when I flew back to the mainland.

I knew I had touched into a power far beyond the worries of everyday life. That it would be there for me each day. That it was real. That I was as much a part of it as the ant or the canyon. And that I would be guided and protected as I moved into my future. That experience changed my life.

Peak Experiences Provide a Transformative Perspective

We've all had a peak experience at some point in our lives. Like Dawn Woodwardson, who told the previous story, it may have happened in nature. It might have happened in sleep. Perhaps you dreamt a profound dream that left you with a feeling of wellbeing and a sense of perspective. Perhaps your peak experience happened when you were playing an instrument, or meditating, or attending a sacred ceremony, or dancing, or exercising, or gazing upon an artistic masterpiece, or daydreaming.[5]

There are many paths to peak experiences. Research shows that virtually every human being has had them, some many times over.[6] Stories from the past, those of shamans from prehistory who "traveled between the worlds" and mystics who described divine rapture, as well as contemporary accounts—by athletes, artists, yogis, scientists, and parents—show them to be a fundamental part of human experience.[7]

They're so profound they often change us forever. At the very least, they make us aware of a sphere of experience that is far different from our everyday reality. At most, they can change our lives completely and immediately.[8]

When perceived through the lens of Spiritual Intelligence, our lives look different. Before her epiphany, Dawn was enmeshed in her suffering. Her mind was hypnotized by problems in all the key domains of her life.

Her peak experience *did not produce a single substantive change* in any of them. Her marriage was still in ruins, as were her business and her health. Nothing had improved.

Yet seen through spiritually intelligent eyes, her entire past had meaning and purpose. She felt differently about it, even though her circumstances were unchanged.

4.2. The perspective from which we see things makes all the difference. We can bless the thorn bush for producing gorgeous roses. Or we can curse the rose bush for producing thorns.

It's not just what happens to us that counts; it's the perspective from which we view it. Spiritual Intelligence gives us a perspective that transforms our understanding of our lives. As we will see in this chapter, it's *the one thing that changes everything else.*

Fundamental Wellbeing and Extraordinary Happiness

After such life-changing experiences, we feel that sense of "fundamental wellbeing" or "extraordinary happiness" deep inside our bones. It has nothing to do with outer circumstances or material security.

During an encounter group at a conference, I was paired up with "Henry," a hedge fund manager. The group leader gave us the exercise of sharing our deepest fears with our partner. Henry confessed to me that his deepest fear was that he would lose all the money in his fund and die in disgrace.

This made little sense because Henry had been honored as one of the most successful fund managers in the industry, year after year, for many years. He was clearly very good at his job, and unlikely to lose his Midas touch. Yet his Caveman Brain, still in survival mode, was forever looking into the shadows, seeking imaginary tigers that might be ready to pounce. His material wealth had not bought him peace of mind.

Monks like Mingyur Rinpoche whom we met in Chapter 2 don't own anything. No house, no car, no pension plan. They don't even own the saffron robes on their backs; technically speaking, their clothing is owned by the monastery. Yet they're living in that state of Bliss Brain every day.[9]

The Inner Experience of Spiritual Intelligence

In this chapter, we're going to examine the specifics of that state. Now that we've taken a tour of the brain and understand the neurological substrate of SQ, we'll focus on the mind and heart. There are several studies of the *experience* of people in Bliss Brain, and these give us a detailed understanding of exactly what they're experiencing. We also draw on the rich tradition of spiritual writing going back thousands of years, based on oral accounts from before the dawn of written text.

You'll find this knowledge incredibly useful as you examine your own life. Modern research, set side by side with snapshots from the minds of the ancient masters, will show you the characteristics, markers, and milestones of SQ and illuminate your own path to Bliss Brain.

Many Roads, One Destination

When you attend a Hindu puja, you notice little in common with a Catholic mass or a Native American pipe ceremony. From the outside, religions look very different. Some believe in one God, others thousands of gods, and yet others no god at all.

Yet behind their wildly different trappings, all religions spring from the same place: the awakened mystical experience.

When he turned 30, Jesus went into the wilderness for 40 days of fasting and prayer, then emerged to begin his public ministry. Buddha spent six years in austere practices before his awakening under the Bodhi tree. Hildegard of Bingen went into trances in which she downloaded prophetic visions of the Holy Trinity. Moses went up Mount Sinai and came back with the Ten Commandments. During Mohammed's night journey, he traveled to the Seven Heavens.

4.3. The transcendent states of mystics are similar regardless of the spiritual tradition in which they originate.

When these teachers came back down to earth, they sought to guide others to that ecstatic state. Their explanations and practices were rooted in the languages, cosmologies, assumptions, beliefs, and cultures of their time, resulting in religions that look very different on the outside.

Yet the mystical experience is the same. Mystics describe becoming "one with everything" and this is the core of SQ regardless of which historical, linguistic, or cultural trappings disguise it.

Through the mystical experience, human beings participate in SQ. We leave our local mind and local reality for a time, and merge with nonlocal mind and nonlocal reality. This awareness is called "nondual" because there is no longer a separation between the two.

As we've seen in the previous three chapters, this isn't merely a subjective experience. It is a neurological event. Activity in parts of the brain drops precipitously. These include the parietal lobe, which locates us in time and space, and the medial prefrontal cortex, which constructs our sense of self.

Our consciousness then escapes the boundaries of ordinary reality, the self bound in time and space, and merges with the field of universal intelligence. These mystical experiences change the brain and the direction of our lives, reinforcing SQ.

Over the past half-century, researchers have applied the tools of science to studying the mystical experience. Investigators have mapped its characteristics and identified the commonalities and differences that define it. This is giving us a clear picture of the awakened experience itself, as well as the brain activity that accompanies it.

The Five Characteristics of Enlightenment Experiences

A key series of research projects began around the turn of the century led by neuroscientist Andrew Newberg, PhD, coauthor of *How Enlightenment Changes Your Brain* and several other books.[10] Andrew put many people having transcendent experiences into fMRI machines and mapped the extensive changes that occur in their brains. He also set up a web portal where people could share their stories of awakening and, in time, he accumulated some 2,000 accounts.

Analyzing these reports, he looked for common themes. The people themselves could not have been more different. Some were Christians, others Muslims, others Buddhists. Some were Hindus, others Jews, others Taoists. Many were atheists or agnostics. They hailed from all over the world. Their accounts differed in the details. However, thematic analysis turned up five common characteristics. These are:

- A sense of **oneness** with the universe and nature
- A sense of **clarity** about the trajectory of one's life
- A perception that their mystical experiences are **more real than everyday reality**
- A feeling of **surrender** to the universe
- The conviction that life has **meaning and purpose**

The reason this research is so exciting is that it associates *subjective experiences* with *objective changes* in the brain. We know when the brain activates in precise ways that this correlates with enlightenment experiences and that those feelings of awakening are correlated with alterations in neural information flow. It's not just the *software of experience* that's changing, it's the *hardware of neurons.* This research shows us that SQ isn't an abstract mental or spiritual phenomenon; it's based on the physiological structure of our body's control network, the nervous system.

Another reason I was so struck by Andrew's research is that the five characteristics are simple and intuitive. So with Andrew and other colleagues, I developed a scale based on them. It's called the Transcendent Experiences Scale. It has now been used in several scientific studies.[11-12] With just five questions based on these criteria, we can determine how close participants have come to an awakened experience. We can also test interventions like EcoMeditation and EFT tapping to find out if they boost SQ, turning Seekers into Finders.

The Four Locations into Which Finders Jump

Another piece of the puzzle was identified by Jeffery Martin. He conducted a large-scale study of people who had reached the tipping point and transitioned from Seeker to Finder. He interviewed over 2,000 people about what their experience of enlightenment felt like.

The astonishing discovery he made was that they didn't all end up in the same place. There was more than one state of enlightenment. In fact, when they made the jump, they landed in one of four places. Jeffery calls these "Locations."[13]

Before Jeffery's work, most people had assumed that one person's enlightenment experience looked much like the next person's. Jeffery's studies showed this was not true. Though there are foundational similarities, there are also distinct differences. Location 3, for instance, looks quite unlike Location 4. And the experience of Location 4 is vastly different in key aspects from Locations 1, 2, and 3. All four Locations have common characteristics, like that sense of "fundamental wellbeing" and the five transcendent experiences that Andrew Newberg profiled. But they differ in essential ways.[14]

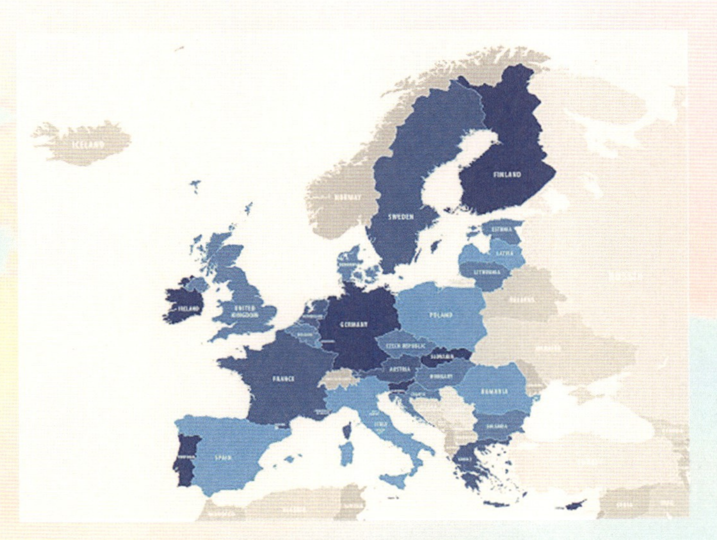

4.4. Countries of the European Union.

It's analogous to a tourist visiting Europe. They're going to use a common currency, the Euro, even though they'll travel to different countries. Many other items are common in the Euro zone, from food labeling standards, to freedom of movement across borders, to the metric system.

But the countries they travel to will differ greatly. One will use the German language, another Spanish, another French, another Danish. One will have an economy based on heavy industry, another on farming. Scandinavian countries have long icy winters while Mediterranean countries bask in warm sunlight more than 300 days each year.

Yet these countries are all still part of Europe.

Before Jeffery's discovery, people in Location 4, for example, thought that Location 4 was "the" experience of enlightenment. And people in Location 3, which differs radically from Location 4, assumed that Location was what enlightenment looked like, and no other.[15]

It's like a person who went to Spain believing that Spain represented Europe, and a person who traveled only to Denmark thought that Denmark was what all of Europe was like. When they talked, they'd describe very different experiences, not realizing that both places were equally valid.

In the same way, if you put a Location 3 Finder in a room with a Location 4 Finder, they will describe experiences as different as Denmark and Spain. Yet both Locations are equally valid places into which people jump once they reach the tipping point.

I have interviewed Jeffery on several occasions, as well as watching his presentations, attending his conferences, reading his book, taking his basic course, and having a series of personal dialogs with him about these states.[16] Below is a summary of the four basic Locations drawn from his book, as well as my own terms that describe each one.

Location 1 – Quieting the Self

1. **Fundamental Wellbeing.** Location 1 Finders have a newfound sense that everything is fundamentally fine. It is usually in the background but occasionally comes to the foreground. It might be disturbed by external events such as the death of a loved one, job challenges, relationship or financial upheaval. However, it returns or can usually be sensed despite the setback.
2. **Mind Chatter.** Finders experience a reduction in the mind chatter of the Narrative Self. Your "inner commentator" quiets down. A sense of spaciousness or inner stillness is felt.

3. **Present-Centered Awareness.** Rather than past and future, Location 1 Finders have greater focus on the present. Memories arise less frequently. Location 1 Finders have less attachment to the story of their lives.
4. **Indifference to Stories.** Finders notice a reduction in interest in all stories. This manifests as less interest in the news, TV series, and the dramas of people around you.
5. **Reduction in Negative Emotions.** A range of positive and negative emotions is still experienced, but negative ones are transient and lack impact.
6. **Self Beyond the Physical Body.** The sense of self of Finders becomes larger and expands beyond the physical body. A feeling of connection between the external and your internal world is experienced.
7. **Reduced Attachment to Outcomes.** Goals can change. Finders have a sense that everything is perfect just the way it is, so they have less attachment to goals. For a period of a few months to two years, they have reduced motivation.
8. **Pain.** Location 1 Finders have a reduced experience of physical pain.

Location 2 – Positive Emotion

1. **Fundamental Wellbeing.** Fundamental wellbeing moves further into the foreground the further one progresses in Location 2. It eventually infuses experience most of the time.
2. **Mind Chatter.** Location 2 Finders see a further reduction in the mind chatter of the narrative self, as well as a reduction of the emotional content of those thoughts that remain.
3. **Present-Centered Awareness.** Deeper immersion in the present moment is a characteristic of Location 2 Finders.
4. **Dissolution of Conditioning.** This Location produces a reduced need for approval by others and for acting in ways that elicit approval.

5. **Increase in Positive Emotions.** Negative emotions are less able to draw Location 2 Finders in; they have diminished emotional reactivity. Emotions are primarily positive.
6. **Dissolution of Boundaries.** The boundaries between "me" and "other" disappear; Location 2 Finders have the classic spiritual experience of "nonduality."
7. **Intuitive Inner Sense of Direction.** Location 2 Finders make intuitively based decisions when faced with choices. Actions seem like part of a great universal order which is moving human beings.
8. **Pain.** This Location produces further reductions in the experience of physical pain.

4.5. The boundaries between self and not-self disappear.

Location 3 – Radiant Love

1. **Fundamental Wellbeing.** For Finders in Location 3, fundamental wellbeing is not only in the foreground but radiates out all around them.

2. **Mind Chatter.** Location 3 produces an even greater reduction in the mind chatter of the narrative self, though it may be noticed more. While suffering is almost completely absent, that which remains is noticeable. Anything that disturbs fundamental wellbeing is difficult to ignore.
3. **Peace and Peak Wellbeing.** Location 3 Finders often believe this is the peak of wellbeing that humans are capable of experiencing.
4. **Profound Sense of Truth.** Finders in Location 3 have a sense of certainty that this is the ideal state.
5. **Positive Meta-Emotion.** One dominant meta-emotion remains. It is a mixture of compassion, joy, love, gratitude, and awe. It is a near-constant experience. Love is felt as divine, universal, and impersonal.

4.6. Location 3 Finders experience universal yet impersonal love.

6. **Union and Connectedness.** This Location produces a sense of oneness with the All That Is. That's the pinnacle experience in many religions.
7. **Service and Outcomes.** While having little need of approval from others, people in Location 3 often engage in service or facilitate others' awakening. However, they are detached from the outcomes.
8. **Pain.** This Location produces even further reductions in the experience of physical pain.

Location 4 – Unfolding with the Universe

1. **Peace and Wellbeing.** In Location 4, peace and wellbeing are an order of magnitude greater than in other Locations.
2. **Mind Chatter.** The narrative self has entirely or almost entirely disappeared. Finders in Location 4 often describe it as "freedom."
3. **Present Moment.** Constant and unwavering immersion in the present moment is a characteristic of Location 4.
4. **Disappearance of Agency.** These Finders have no sense of agency, or the ability to make decisions. Life is unfolding and they watch it happen.
5. **Disappearance of Emotion.** The experience of emotion disappears in this Location.
6. **Nonduality.** There is no sense of a separate self. The universe, speaker, hearer, are all one.
7. **Memory Deficits.** Location 4 Finders often have an inability to recall scheduled events. Memories aren't being stored in their minds, especially those to do with stories.
8. **Approval.** This Location is characterized by a complete disappearance of need for approval from others.

4.7. Location 4 Finders no longer pay attention to what other people think they should be doing.

As you read this list of characteristics, you might find some familiar. You may have touched one or more of these four Locations at some point in your life. Perhaps you haven't reached the tipping point of where you're in fundamental wellbeing all the time, but you've at least tasted the ecstasy of a Location at one time. In the Extended Play Resources at the end of this chapter, you'll find Jeffery's test for determining which Location you're in.

FROM STRIPPER AND DRUG DEALER TO SPIRITUAL TEACHER

Mandy Morris's life story is one of extraordinary transformation, marked by extreme highs and lows. Her early years were fraught with disappointment and hardship. Growing up in an unstable environment, Mandy experienced poverty, neglect, and abuse. By the age of

18, she wound up working as a stripper and dealing drugs to make ends meet. "I felt like I was trapped in a cycle of darkness, with no way out," she recalls.

During this turbulent time, Mandy worked in a junkyard by day and navigated the dangerous world of drugs and sex by night. The constant struggle for survival took a toll on her spirit. Yet, amid the chaos, she felt a flicker of hope inside. "There was a small voice inside me that kept whispering, 'This isn't all there is for you. There is more to life than this.'"

Her turning point came when she met a spiritual mentor named Daniel, who introduced her to the concepts of manifesting and the power of the mind. Daniel saw potential in Mandy that she couldn't see in herself. "Daniel told me that I had the power to change my reality through my thoughts and intentions. It was a concept that seemed foreign to me, but I was desperate for a change."

For Mandy, the tipping point came one night while meditating with Daniel: "I saw myself surrounded by light, free from all the pain and suffering of my past. It was as if a veil had been lifted, and I could see the truth of who I really was. I realized that my past did not define my future and that I had the power to create the life I wanted."

Once she'd passed the tipping point, Mandy dedicated herself to learning and practicing the principles of manifesting. Her journey of self-discovery and transformation led her to write the book *8 Secrets to Powerful Manifesting,* where she shares the insights and techniques that helped her turn her life around.[17]

Today, Mandy Morris is a celebrated author and spiritual teacher, inspiring countless others to harness their inner strength and create a life of their dreams. Her story is a powerful reminder that no matter how dark the past, the future holds limitless possibilities.

As you can see from Mandy's story, once the tipping point is reached and one enters that state of extraordinary happiness, all of life changes. In Mandy's words, the veil is lifted. Every domain of your life is transformed by SQ. You discover your inner purpose and begin living a purpose-filled outer life. Your inner life becomes radiant.

As SQ rises, even your body and mind become radiant. When you walk up to a spiritual master, you can sometimes feel that radiance. The light within them lights up the whole room. They don't need to be an official guru or teacher. There are "ordinary" people whose inner luminosity is so great that it shines all around them. This is common to many people who are in Location 3.

The Positive Meta-Emotion

That "single positive meta-emotion" is a particularly interesting finding of Jeffery's study. MRI studies typically characterize the positive emotion felt by master meditators as "compassion."[18] That label comes from Buddhism and has migrated into brain research. Investigators ask monks to meditate on "compassion" and measure the changes in their brains when they move into this emotional perspective.

4.8. Radiance begins to shine all around people who live in Location 3.

However, when you experience this state yourself, you'll discover that it's much more than compassion. It's a glorious combination of awe, joy, compassion, gratitude, and ecstasy. This is characteristic #5 of Location 3. Jeffery Martin coined the term "a single positive meta-emotion" to describe it.

It's a positive feeling so intense that it seems to burst out of your heart, fill your body and mind, and radiate throughout the space all around you. You may cry, laugh, or shake because the feeling is so intense. You can't describe it as any one emotion because it's every positive emotion at the same time. When you're feeling this blissful, you have no interest in naming things or breaking down the components of this overwhelming exquisite sensation. You're swept away beyond the definitions of the mind.

Patanjali and the Four Basic States of Samadhi

Fundamental wellbeing, aka extraordinary happiness, is a characteristic common to all four of the Locations identified by Jeffery Martin. Ramana Maharshi and other great spiritual masters have also affirmed that regardless of which spiritual path you take, bliss is the common characteristic of all the varieties of the enlightened state.[19]

We ended Chapter 2 with an exploration of the study of Alzheimer's patients. It found that, rather than shrinking as in the normal course of the disease, meditation triggered growth in important brain regions related to executive control, emotion regulation, memory, and learning.

Study participants used a meditation based on the *Yoga Sutras of Patanjali*. This classic scripture was first committed to writing around 2,000 years ago but is based on an oral tradition that is thousands of years older. That oral tradition represents the cumulative spiritual knowledge of the entire Vedic civilization of India.

The word "yoga" as used in this philosophy does not mean the asanas or physical poses assumed by people taking a Hatha Yoga class. It means "union" in the sense of the dissolution of the boundaries between universal consciousness and the individual consciousness of a human being. "Yoga" is oneness and nonduality. A "yogi" is a person who has learned to release everything between herself and infinite consciousness, and entered a state of union with All That Is.

Sutras and Yoga Philosophy

The sutra form is a distinctive form of literature. Each sutra is just one single sentence long. Yet it is written to embody a profound truth. Every word is chosen with the utmost care. Patanjali's Yoga Sutras contain just 195 one-liners, yet these constitute a comprehensive summary of the entire path to spiritual awakening and the essence of SQ.[20]

According to Patanjali, the ultimate goal of meditation and other spiritual practices is *Samadhi,* and many sections of the *Yoga Sutras* cover various aspects of how to attain it. The word *Samadhi* has many nuances and connotations. These include complete focused absorption, bliss, and union with cosmic consciousness. Yogis experience "extraordinary happiness" as their minds are flooded with that "single positive meta-emotion" when in Samadhi.

The initial stages of Samadhi involve detaching the mind's attention from the outer world and turning it inward. In the ultimate stages of Samadhi, distinctions between self and the infinite disappear. You experience union with—or absorption into—ultimate reality. In this state, all separation between self and the infinite dissolves.

4.9. Meditator entering a state of Samadhi, blissful union with infinite consciousness.

The yogis summarize the experience in three words: *sat, chit,* and *ananda*. Those are the Sanskrit words for "being," "consciousness," and "bliss."

Vedantic philosophy holds that every human being is designed to live in this state of perfect undying bliss. Rooted in being, our individual human consciousness is destined to lose itself in oneness with the great universal consciousness of the universe. It further teaches that every single human being will eventually liberate themselves from the obstacles that stand between us and this state.

The time frame in which we do this is largely up to us. Some people will structure their lives around attaining it and do so in days, months, or years. Others will put it off for decades. Some people won't bother seeking the blissful state that is their birthright at all. They may go from birth to death without experiencing more than a few accidental moments of it. According to another Vedic sage, Shankara, this is the missed opportunity of a lifetime.[21]

For people who make no space in their lives for SQ, liberation from the obstacles to bliss may take many lifetimes, even thousands of lifetimes. But since you and I have the option of living in bliss now or putting it off for centuries, why not devote our full attention to it now?

The Modifications of the Mind

In Sutra 1.2, Patanjali describes the goal of spiritual practice as being to restrain the "modifications of the mind." Vedantic philosophy teaches that our unmodified state is bliss. But our minds get in the way. They do this by wandering. They flit from thought to thought, distracting us from our natural blissful state.

The practices of Vedanta train us to restrain the mind's wandering tendencies. They teach us to detach the mind from focus on the outside world and direct its attention inward. We then find Samadhi.[22]

The four circuits of the Enlightenment Network, working together, restrain the "modifications of the mind" allowing us to perceive the actual unmodified reality of universal consciousness. The Emotion Regulation Circuit dampens distracting feelings. The Attention Circuit inhibits the mind's wandering tendencies by focusing it on a single object. The Selfing Control Circuit quiets the inner chatterbox that distracts us from the present moment. This gives us access to the bliss to be experienced as the Empathy Circuit lights up and we feel that "single positive meta-emotion."

In this way, just as they did with the Alzheimer's patients of Chapter 2, Patanjali's practices detach our minds from their obsessive focus on the outside world with all its distractions. They shift the attention of the mind inward, to the place we experience Samadhi.

In the *Yoga Sutras,* Patanjali describes four initial states of Samadhi. The first is easily achievable by even the novice meditator. The states lead one after the other to progressively greater release of the "modifications of the mind" and set us on the path to ecstasy.[23]

These four basic forms of Samadhi are closely aligned with the Four Locations described by Jeffery Martin. You're likely to recognize these states because you may have experienced one or more of them, even if you've only been a casual or occasional meditator.

Vitarka Samadhi

To assist beginning meditators in withdrawing our attention from the outer world and developing single-minded concentration, Patanjali has us choose a concrete physical icon to guide us.

That's because we begin our journey to SQ while embedded in the material world. It's easier for ordinary human beings starting out on their journey to focus on a familiar material object, rather than attempting to direct their attention to the immaterial and nonphysical targets used by accomplished yogis in later stages of Samadhi.

A physical target can take many forms. It might be the image of a saint, the echo of a mantra, a profound verse of scripture, a candle, a piece of artwork, a sublime visualization, or the rhythm of our own breath. This anchor in the physical realm helps us concentrate, a starting point for our inward exploration, and takes us to the first stage, Vitarka Samadhi.

In Vitarka Samadhi, our attention is absorbed in the physical object. Every aspect of the object is observed to the point where our attention is fully absorbed in the physical qualities of the object.

As you fire up your Attention Circuit, this shifts your focus away from your "ordinary" self, with its worries, frustrations, judgments and preoccupation with past and future. The Selfing Control Circuit becomes active and you're no longer obsessed with the past and future. You enter the present moment, available for the stillness that comes with Samadhi.

You feel the "fundamental wellbeing" of Location 1 as mind chatter reduces and negative emotions drop off. You experience other Location 1 characteristics like feeling your sense of self expanding beyond your skin sack, reduced attachment to your goals, and indifference to the "stories" out of which we build our lives.

4.10. In Tibetan Buddhism, mandalas are often used as an object of contemplation.

Vitarka Samadhi achieves this by shifting the meditator's attention from enmeshment with the ordinary self—constructed by the medial prefrontal cortex—to absorption in an inspirational physical object that represents the sublime.

Vichara Samadhi

In the next phase of meditation, we transition into the Vichara stage of Samadhi. Here your attention moves beyond the outer physical aspects of the object to its nonphysical subtleties.

In Vichara Samadhi, you become aware of its inner reality, of finer details, intricate patterns, and deeper truths about the object. Abstract qualities such as beauty, love, color, sound, texture, or the flavor of the object are discerned. In many EcoMeditation tracks, like the one you'll find at the end of this chapter, I guide you into exploring all the sensory qualities of an object. This is because taste, touch, smell, hearing, and sight all use different areas of the brain. When you use all five senses, you recruit many different regions of the brain and focus them all in the same direction.

Vichara Samadhi requires a greater degree of focus than Vitarka Samadhi. That's because it's harder to focus on a nonphysical target that you can't observe via your senses.

This is where the orbitofrontal cortex, a key component of the Attention Circuit, plays a part. It's highly active in experienced meditators. As the Selfing Control Circuit dials down the commentary of your inner chatterbox, the Attention Circuit activates fully and you're able to focus on the nonphysical aspects of your object.

If you're using a mantra as your object of meditation, in Vitarka Samadhi you're focused on the words. When you move to Vichara Samadhi, you might discern the inner essence of the mantra's meaning. If you're focused on your breath in Vitarka Samadhi, you might meditate on the life force flowing through your breath in Vichara Samadhi. All the positive qualities of experience intensify, just as they do for Location 2 Finders.

With practice, entering this elevated state becomes easier. Clinging to the physical world diminishes, as well as to the self that's embedded in it, and the mind moves inward from Vitarka Samadhi to Vichara Samadhi.

4.11. Buddhist monks in Thien Mu Pagoda, Vietnam, chanting traditional prayers.

Ananda Samadhi

The next stage on the journey is Ananda Samadhi. At this point, we've moved beyond the tangible physical object of meditation, as well as its subtle qualities. We're enveloped in a state of pure joy. Bliss is our object of focus.

In Ananda Samadhi, the cognitive parts of the brain dial down dramatically. The meditator's attention releases reflection and reasoning, and settles into blissful tranquility. The mind floats in a sea of undifferentiated joy. It has withdrawn from enmeshment with the outside world, focused itself inward, and awakened its natural state of ecstasy.

This is reflected in brain activity. The inner commentator of the ordinary self is now quiet. We see a further dropoff in the activity of the medial prefrontal cortex, which constructs that illusory self. In one of my key MRI studies, we also observed the busiest of the executive centers, the dorsolateral prefrontal cortex of the brain's left hemisphere, shutting down.[24]

The insula and other parts of the brain that mediate positive emotions like compassion, awe, gratitude, and bliss are highly active.[25] Our bodies might tremble or shake with orgasmic ecstasy as we feel that single overwhelming positive meta-emotion. Ananda Samadhi bears a close relationship to Location 3.

We lose interest in obsessive negative thinking when we're enveloped in ecstasy. Sigmund Freud coined the term "oceanic bliss" which is an apt description of the experience of Ananda Samadhi. Bliss is the object of our focus in this third rung on the ladder of Samadhi.

Asmita Samadhi

In Asmita Samadhi, emotion fades into the background and there is simply "I am." We become observers, detached and serene, in the witness state of I-ness. We've used our journey with the object of meditation to bring us to awareness, then to bliss, and then to the fundamental ground of being. Our own I-ness is the object of concentration. We rest in SQ.

In Asmita Samadhi, we're fully in the present moment. There is no ego, fear, or desire. The mind is witness to the material world from the highest vantage point. It knows that "I" am not my experiences, thoughts, or emotions, but the awareness observing them. The mind knows itself to be an expression of the eternal consciousness that underlies everything. Asmita Samadhi closely parallels the experience of Finders in Location 4.

In this state, the brain is quiet. The breath is slow and even. The body is still and calm. In Asmita Samadhi, we know ourselves to be the consciousness of the universe looking out through human eyes.

Nonduality

A consistent report of Finders in these ecstatic Locations, as well as the goal of SQ, is a *sense of oneness*. They feel one with nature, with all other life forms, and with the universe itself. They don't perceive "me" over here and "other" over there. "Nonduality" is an apt description of this dissolution of the individual into the consciousness of the whole of creation.

This perspective is most intense in Location 4 Finders. Like the "I am" state of Asmita Samadhi, it dissolves all boundaries between self and other. The individual identity merges with universal consciousness. In the Transcendent Experiences Scale I developed based on Andrew Newberg's book *How Enlightenment Changes Your Brain,* nonduality is the first and defining characteristic: "A sense of oneness with the universe and nature."

Nondualism refers to the concept that there is a single unitary consciousness and that human consciousness can merge with it. Nondualism transcends all perceptions of "I" and "other."

This "unity consciousness" is a central experience of Finders. There are many teachers and teachings about these states of "I-am-ness" today. Most of them take their inspiration from a great Indian sage of the past century, Ramana Maharshi. His awakening to the nondual state happened when he was just 16 years old and has been an illustration of SQ for generations of Seekers.

RAMANA MAHARSHI, THE GREAT TEACHER OF NONDUALITY

Ramana Maharshi, born Venkataraman Iyer in 1879, was one of the most revered Indian sages of the 20th century. He was renowned for his simple, direct path of self-inquiry as a tool for spiritual awakening. His enlightenment experience, a profound, transformative event that occurred spontaneously during his teenage years, is a cornerstone of his teachings and spiritual legacy.

In 1896, Venkataraman's father died and his family went to live with his uncle in Madurai, Tamil Nadu.

One day in mid-July, sitting alone in his uncle's house, he was suddenly stricken with an intense fear of death. This was not triggered by any illness or external threat but arose from within, bringing him face-to-face with the fundamental fear of mortality.

Unlike most of us, the boy did not flee this fear; instead, he decided to probe it directly. In a state of deep contemplation, he lay down, turned his attention inward, and inhabited the experience of being a corpse. He remained motionless, holding his breath, focusing on the question of who it is that dies. This inquiry led him to a profound realization: it was not his body that constituted his true self, but something eternal and immutable.

Venkataraman's fear dissipated, replaced by an overwhelming sense of awareness. He realized that death pertains only to the body and that the true "I," or Self, is deathless, existing beyond physical demise. This spontaneous awakening was not an intellectual understanding but a direct, experiential insight into the nature of his true Self. From that moment, Venkataraman's outlook on life was irrevocably altered. He saw that the essence of his being, the Self, was eternal and one with the source of all existence. This realization is central to yogic philosophy, which teaches that the individual soul is one with the infinite universe.

4.12. Ramana Maharshi.

Following this experience, Venkataraman lost interest in worldly affairs, school, and his family. He sought solitude and silence to immerse himself in the bliss of Self-awareness. Shortly afterward, he left for the sacred mountain of Arunachala in Tiruvannamalai, drawn by a mystical attraction to the place he had only read about in a sacred text. This marked the beginning of his life as a sage.

Ramana Maharshi's teachings revolved around the direct path of self-inquiry. He urged Seekers to constantly inquire "Who am I?" to penetrate the illusions of the ego and realize the Self that is one with everything, the essence of nonduality.

Unlike other paths that may involve complex rituals or intellectual understanding, Ramana's approach was remarkably simple yet profoundly effective. He taught that the realization of the Self could be attained here and now by anyone willing to earnestly engage in this inquiry, leading to liberation from the cycle of birth and death.

He welcomed many visitors to the ashram that grew up around him. Among those was Paul Brunton, who wrote about him in his 1932 book *A Search in Secret India*.[26] The book sold more than a million copies, introducing Ramana Maharshi—and nondualism—to the world.

Ramana Maharshi's enlightenment experience exemplifies the transformative power of facing one's fears, looking beyond material experience, and seeking union with the infinite. His life and teachings have inspired many of today's spiritual teachers and his ideas continue to resonate a century after they were popularized by Paul Brunton.

Ancient and Modern Perspectives

It's remarkable that the brain regions of the Enlightenment Network can now be tied to particular types of Samadhi described in a 2,000-year-old scripture, to the nondual experience exemplified by Ramana Maharshi, and to modern research into the Finders. Today's scientists are enriching our understanding of states that have been experienced by spiritually intelligent sages for thousands of years.

It's equally astonishing that when Alzheimer's patients use meditative techniques based on Patanjali, they literally grow the memory and learning centers of their brains. It's not just happening to people with cognitive decline; when you and I practice these methods, they grow those parts of our brains too.[27] Developing SQ gives us access to the *sat, chit,* and *ananda* of the masters—the "being, consciousness, and bliss" of the nondual awakened state.

While there are several flavors of the awakened experience, as we see from both the Finders study and Patanjali's stages of Samadhi, they have common central characteristics. One of these is mind-blowing ecstasy.

Ramana Maharshi tells us that there are many paths to the infinite, but ananda is common to all of them. Bliss is a hallmark of awakening.

While in Chapters 1 through 3 we examined the role the brain plays in creating and developing SQ, this chapter has given us a detailed view of what the inner personal perspective looks like. We now understand the experience of the mystic and how it has common characteristics, whether these are described in ancient scriptures or modern neuroscience.

In Chapter 5, we'll see how nondual and ecstatic experiences aren't the exclusive property of mystics. Everyone has them and ordinary "non-spiritual" people can experience them too. That's the magic of "flow" and, as we tap these states, we incorporate the ecstasy of ananda into our daily lives.

Deepening Practices

The Deepening Practices for this chapter include:

- **Journaling Exercise 1:** As you read through the Locations and the types of Samadhi, which ones might you have experienced, even briefly? You don't have to be sure, just take your best guess. Record the experience in your journal.
- **Journaling Exercise 2:** Who is a person you've known who simply radiated love? Record their name in your journal and describe how you felt when you were around them.
- **Dawson Guided EcoMeditation:** Sensory Aspects of Experience

Get the meditation track at: SpiritualIntelligence.info/4.

Extended Play Resources

The Extended Play version of this chapter includes:

- Jeffery Martin's test for which Location you're in
- Swami Satchidananda's description of the stages of Samadhi
- Chopra Center blog on the 3 levels of Samadhi
- Podcast: The Finders: Jeffery Martin and Dawson Church in Conversation

Get the extended play resources at: SpiritualIntelligence.info/4.

Updates to This Chapter

As new studies are published, this chapter is regularly updated. Get the most recent version at: SpiritualIntelligence.info/4.

Chapter 5

Enlightenment Without Spirituality

The Brain Waves of Spiritual Intelligence

Among the most brilliant and innovative trailblazers in the field of Spiritual Intelligence, Maxwell Cade ("Max" to his friends) stands out.

A distinguished biophysicist, Max made key contributions to the development of radar technology for the British government during World War II. After the war, he focused his immense scientific gifts on exploring elevated states of consciousness.

He became friends with many of the great spiritual teachers of the 20th century. Among these were Swami Muktananda, a renowned Indian spiritual leader known for promoting the Siddha Yoga path, and the Zen master D. T. Suzuki, who played a key role in introducing Zen Buddhism to the Western world. Max was also influenced by the teachings of Lama Anagarika Govinda, a German-Tibetan monk and a scholar of Tibetan Buddhism. In addition, Max trained as a master hypnotist.

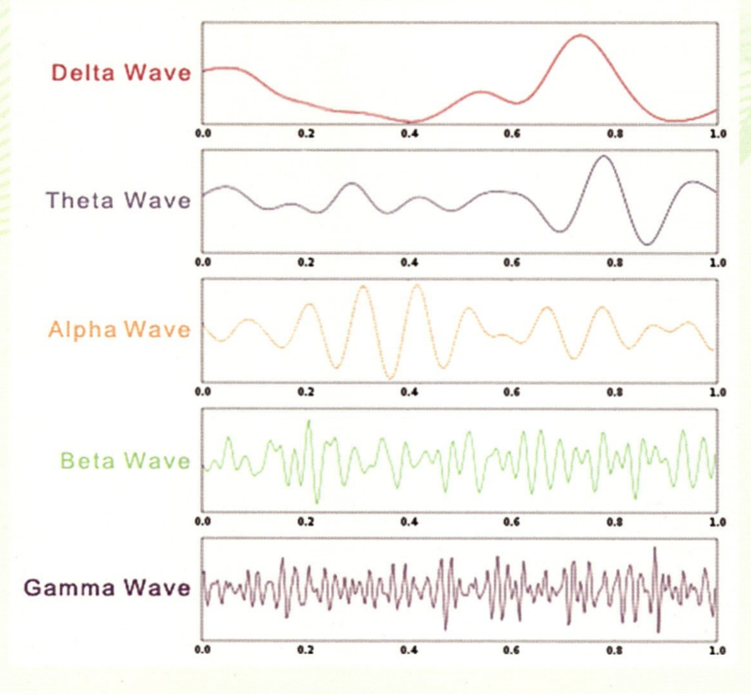

5.1. Brain-wave frequencies from the slowest wave (delta) to the fastest wave (gamma).

He became fascinated by the new field of electroencephalography, which had emerged just before the war, seeing it as a bridge between science and spirituality. Max tinkered with a variety of designs and in 1976 came up with a device he called the Mind Mirror. Unique among electroencephalograms (EEGs), it didn't represent brain waves as squiggly lines on a screen. Instead it used horizontal rows of colored lights, with each row denoting a particular brain-wave frequency.

The Mind Mirror gave both Max and the user a *visual representation* of the frequencies being generated by the brain and the relative strength of each one. It allowed Max to gather data and to train the user to shift their mental activity to produce different brain-wave patterns. The Mind Mirror was one of the first practical scientific tools for people seeking to develop SQ.

Using the Mind Mirror over a span of 20 years, Max recorded the brain-wave patterns of over 4,000 spiritual adepts. He discovered that their brain function altered when they entered elevated states. They all exhibited a common pattern. The *ratio* of the primary waves—from delta, the slowest, to gamma, the fastest—conformed to a certain shape. Max called this the "Awakened Mind" pattern.

Here's how Anna Wise, a protégé of Max's, described his work: "What sets the Mind Mirror apart from other forms of electroencephalography was the interest, on the part of its developer, not in pathological

5.2. Max Cade with a group of students using the first 1970s version of the Mind Mirror.

states (as in the case of medical devices), but in an optimum state called the Awakened Mind. Instead of measuring subjects with problems, the inventor of the Mind Mirror sought the most highly developed and spiritually conscious people he could find. In the flicker of their brain waves, he and his colleagues found a common pattern, whether the subject was a yogi, a Zen master, or a healer."[1]

The Awakened Mind

Anna's observation points to an astonishing feature of Max's discovery. The spiritual and religious beliefs and training of these spiritual adepts could not have been more diverse. They spanned the whole range of traditions, including Indigenous shamans, Christian faith healers, Taoist masters, Jewish kabbalists, Buddhist monks, Native American medicine women, Catholic nuns, Islamic Sufis, and Qigong masters.

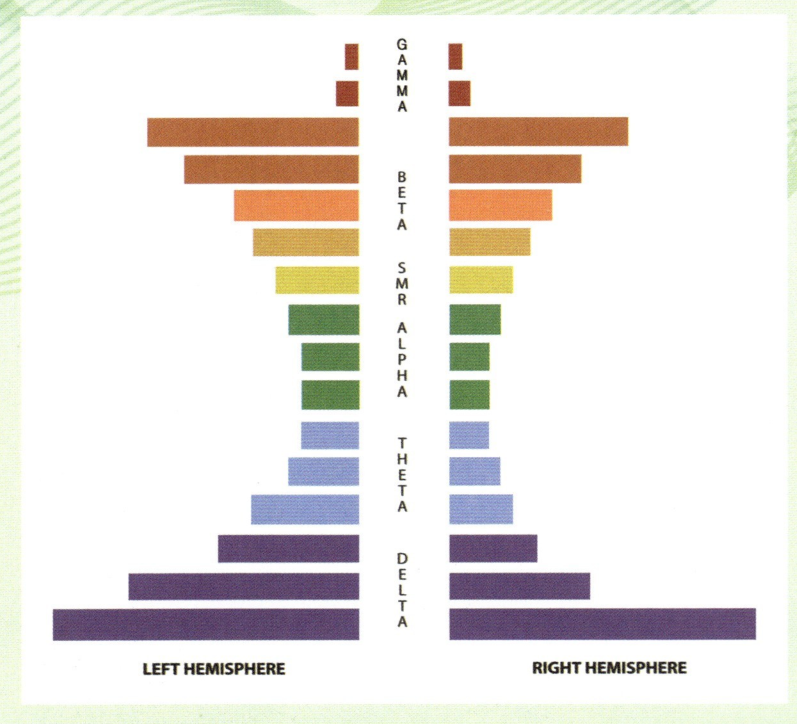

5.3. Brain wave frequencies of ordinary consciousness. Note attenuated alpha and substantial beta.

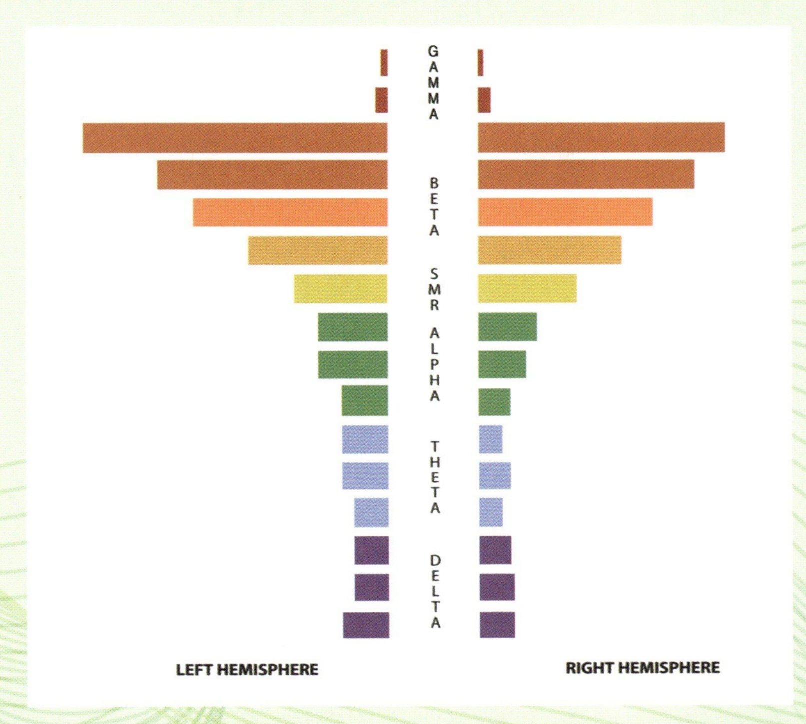

5.4. Stress. Note extensive beta and very little of any other frequency.

5.5. The Awakened Mind pattern. Notice how much larger alpha and gamma are than in ordinary consciousness in 5.3.

Yet when they entered the mystic state of interaction with universal consciousness—which is how we defined SQ in Chapter 1—*their brain waves all looked similar.* The pattern of the Awakened Mind appeared on the Mind Mirror, whatever religion or faith tradition the adept practiced. This made it clear that at the level of brain function, *the experience of the mystics is one,* even though the paths they take to get there are many.

Cade also noted another commonality: These adepts exhibited high levels of alpha waves. Alpha waves are centrally located in the frequency spectrum, with high-frequency beta and gamma above, and low-frequency theta and delta below. In the Awakened Mind state, abundant alpha waves form a bridge linking the high and low frequencies.[2]

Cade termed this the "alpha bridge" because alpha sits right between the high frequencies of conscious and superconscious mind (beta and gamma) and the low frequencies of subconscious and unconscious (theta and delta). This alpha bridge facilitates the exchange of information between the high- and low-frequency bands, leading to heightened perception, intuition, creativity, flow, resilience, and cognition. When hooked up to an EEG, people with high SQ typically show enhanced alpha waves.

You can see from the images above how distinctive the Awakened Mind pattern is and how different it is from both ordinary consciousness and stress. In a conventional state of mind, beta predominates. It's the wave of everyday awareness, and when we're thinking and acting in our ordinary lives, beta uses up most of the brain's available energy. There's not much left over for anything else.

When we go into Caveman Brain, stressed and afraid, beta completely takes over. The other brain waves can disappear completely. People in negative emotional states have a huge amplitude of beta and none of the balancing properties of the other waves.

In the Awakened Mind, we see the development of a wide alpha bridge. Big delta provides a solid base at the bottom of the scale and, on this, the positive emotions of gamma build at the top.

What Brain-Wave Frequencies Mean

Brain waves are measured in amplitude (size) and frequency (speed). They are a useful way of visualizing the energy flows of the brain. Beta is the brain wave of ordinary consciousness and is most predominant when we're awake. But when we get stressed, our brains produce excessive amplitudes of beta.

When we're sleeping, our brains slow down. Most of the night, they're in delta, the slowest of waves, with occasional flares of theta when we dream. During these theta bursts, our brains build new networks in the hippocampus, reinforcing learning and insight. Another situation in which theta predominates is in the brains of healers when they're in the midst of facilitating healing for others.

Gamma waves are associated with the *synchronization of information from all four primary lobes of the brain.* Gamma is found in many positive mental and spiritual states. These include love, insight, compassion, awe, perceptual organization, associative learning, flow, healing, attention, patience, gratitude, and transcendent bliss.[3]

Gamma waves synchronize the firing of the whole brain as a coherent whole. They're found in highly creative people, as well as ordinary people having a moment of insight. Gamma is also observed in states of mystical union. When studying advanced yogis, Richard Davidson found that the gamma activity in their brains increased 25-fold.[4]

Anxious and stressed people produce large amplitudes of beta. Because beta hogs all the brain's energy, there's not much left over for the calming wave of alpha, the healing wave of theta, or the coherent wave of gamma.

Meditators, on the other hand, show greatly reduced beta activity.[5] This makes energy available for the other brain-wave frequencies—from gamma at the top to delta at the bottom. What EEG experts have found is that big gamma "rides" on big delta.[6] We develop high amplitudes of delta first and this forms a base for high amplitudes of gamma. Steven

Kotler, one of the best-known researchers into flow states, calls gamma a "coupled wave" since it grows following the expansion of delta and theta.

As you can see from this very brief review, brain waves are closely associated with mood and SQ. They are so revealing that an EEG expert can look at a person's brain waves and determine immediately what sort of mental state they're in *without knowing anything about them.*

Techniques in biofeedback and neurofeedback emphasize boosting your alpha waves first. This generates sufficient alpha to unify the faster and slower waves using the "alpha bridge" that Max Cade discovered.

EcoMeditation and the Awakened Mind

When I first developed EcoMeditation, I was gifted to have several colleagues trained on the Mind Mirror available to measure the brain waves of participants at my workshops. One of these, Judith Pennington of the Institute for the Awakened Mind, has collected a database of over 10,000 brain-wave scans.

The first of my participants Judith hooked up was a group of eight people attending an EFT workshop in New York City. Before we started training in EFT, I had all the participants practice EcoMeditation for 20 minutes. Two groups of four people each were tethered to the Mind Mirror.

5.6. EEG hookup showing Awakened Mind at my Life Vision Retreat.

During the next break, I asked Judith and her assistant what they had observed. They were beside themselves with excitement. They'd seen the "big alpha" typical of the Awakened Mind form first, then the typical pattern of increased delta and gamma. Some participants had so much delta that Judith and her assistant had to zoom their screens out to capture it all.

Many of these workshop participants weren't even regular meditators. We've had hundreds of reports from people who've tried many other forms of meditation in an attempt to find inner peace, been unsuccessful, and given up. But when they try EcoMeditation, they finally hit the bullseye—often *the very first time.* At the workshop, these novices were amazed that they were finally able to achieve a calm inner state.

One, Maria, had been struggling with depression for several years. After inducing the relaxed state that begins each EcoMeditation session, I used a guided imagery exercise developed by Max Cade. It has participants call on inner allies or archetypes to support their transformative journey. Here's what Maria shared with the researchers after we completed the meditation session.

THE DIAMOND CRYSTAL IN MY THROAT

"At first, having my eyes closed was annoying. I could feel every little scratchy itchy feeling in my skin. My throat tickled, and I wanted to cough. I could hear the guy next to me breathing, and that was annoying too. But then I began to forget about all that stuff, and a feeling of peace came over me.

"I could feel the breath going inside my body. And going out again. It felt like a river flowing. I started to float, like I was a helium balloon or something.

"I seemed to go to another place, and it was beautiful. I could feel the rocks and trees and ocean, and I seemed to be part of it all, like I was absorbed into this perfection of everything there is in the cosmos.

"These four huge blue beings drifted near me, and I felt incredible love and connection flowing out of them. They were like outlines of people but transparent and about 15 feet high. Made out of a beautiful royal blue mist.

"I've been so worried about all the stuff going on in my life lately, but one of the beings drifted close to me and I felt reassured. Like she was telling me everything is going to be okay. My heart filled up with love, and I realized that love is everything.

5.7. The diamond crystal in throat and heart.

"She gave me a shiny diamond crystal to remind me that she's always there for me. I put it in my throat and heart. It melted all the miserable, depressed pain that's been living there for too long, and the pain became drops of water that fell into the ground.

"When you told us to come back into the room, I felt like I was a million miles away. I brought that feeling of peace back into my body. It was hard to come back, and I realize part of me is there all the time."

In these profound states of consciousness, people, like Maria, often discover inner resources they didn't realize they had. In this serene realm, enveloped in deep tranquility, Maria encounters remarkable allies like the ethereal blue beings. They bestow gifts symbolic of profound personal growth, such as the gleaming crystal of transformation.

When you return from "a million miles away," you have a new perspective on your life. Max Cade once wrote: "The awakening of awareness is like gradually awakening from sleep and becoming more and more vividly aware of everyday reality—only it's everyday reality from which we are awakening!"[7]

The Concept of Flow

When they're performing at their best, elite performers—whether they're musicians, athletes, businesspeople, teachers, artists, scientists, accountants, lawyers, warriors, salespeople, or coders—are in that unique brain-body state called flow. Flow is a mental state of deep, sustained focus. When in flow, you experience *optimal engagement, enjoyment, and performance* while doing an activity. Your focus gets so intense that action and awareness start to merge and everything else disappears.

In flow, you're totally immersed in your current activity and the present moment. Flow can arise during a wide variety of tasks like learning, creating, or participating in a sport. When in a flow state, people pay no attention to distractions and time seems to pass without being noticed.

This state is marked by intense concentration, a sense of control, and a loss of self-consciousness. Flow balances the challenges of a task against a person's skills, leading to a deeply rewarding experience. The account of the discovery of the periodic table of the elements by Russian chemist Dmitri Mendeleev in the mid 1800s contains all the ingredients of flow.

THE DREAM THAT LED TO A SCIENTIFIC BREAKTHROUGH

In 1867, Russian professor Dmitri Mendeleev began writing a book called *Principles of Chemistry.* During his classes, he had observed that students had difficulty understanding the chemical elements because there was no clear way in which they were organized and related to each other.

At that time, science knew about the existence of 63 elements. In order to figure out how to organize them, Mendeleev made up a set of playing cards on which he wrote the atomic weight and the properties of each element. He became obsessed with different ways of arranging the cards and took them everywhere he went.

On February 17, 1869, he was due to catch a train soon after breakfast. But he became lost in rearranging the cards in various sequences. He noticed that there were some gaps in the order of atomic mass and tried various combinations to resolve them.

Mendeleev became so absorbed in the process that he not only forgot the train, but continued rearranging the cards for three days and nights. Finally, he fell into an exhausted sleep.

Here's how he described what happened next: "I saw in a dream, a table, where all the elements fell into place as required. Awakening, I immediately wrote it down on a piece of paper."[8] He named his discovery the "periodic table of the elements."

They revealed what came to be called the Periodic Law, because certain atomic properties repeat themselves at periodic intervals.

But this systematic organization left gaps in the chart. This led Mendeleev to predict that there must be additional elements that had not yet been discovered. He extrapolated the chemical properties and atomic masses of the missing elements from the 63 known elements. His colleagues now had clear targets to search for.

Eventually, Mendeleev completed his book. It became the standard text for the field for half a century.

The word "flow" was first given its current meaning by Hungarian-American psychologist Mihaly Csikszentmihalyi. He defined it as "a state in which people are so involved in an activity that nothing else seems to matter; the experience is so enjoyable that people will continue to do it even at great cost, for the sheer sake of doing it."[9]

David Laney, *UltraRunning* magazine's 2015 male ultra-runner of the year and a Nike trail running coach, says, "There are days when you are invincible." When you are in flow, you are "just so confident and calm. There's nothing that you can't do."[10]

Csikszentmihalyi was not the first to identify flow. A generation earlier Abraham Maslow placed "self-actualization" at the very top of his famous pyramid called the "hierarchy of needs."

But he noticed there was a small subset of people who, having attained self-actualization, went on to a further stage of personal development. He first called them "transcending self-actualizers."[11] After self-actualization, they went on to a higher state of consciousness, which Maslow referred to as a "peak experience."

Among the characteristics of this subset of people were the persistent experience of elevated emotions such as awe, appreciation, and gratitude. Maslow noticed that they valued the experience of simply "being" above the activity of "doing."

During the last couple of years of his life, Maslow began to experience such states himself. He concluded that his original hierarchy of needs was incomplete, and placed a new level above self-actualization.[12]

Self-Transcendence

Maslow called this new capstone "self-transcendence" because people in this state transcended the "self" with a small "s" they'd previously believed they were. They interacted with a universal consciousness greater than their human consciousness—the definition of SQ we learned in Chapter 1.

Ironically, the meme of Maslow's old pyramid became embedded into popular culture even as Maslow himself explored and defined the dimensions of personal transformation beyond self-actualization. He was not able to publish his observations prior to his death.[13]

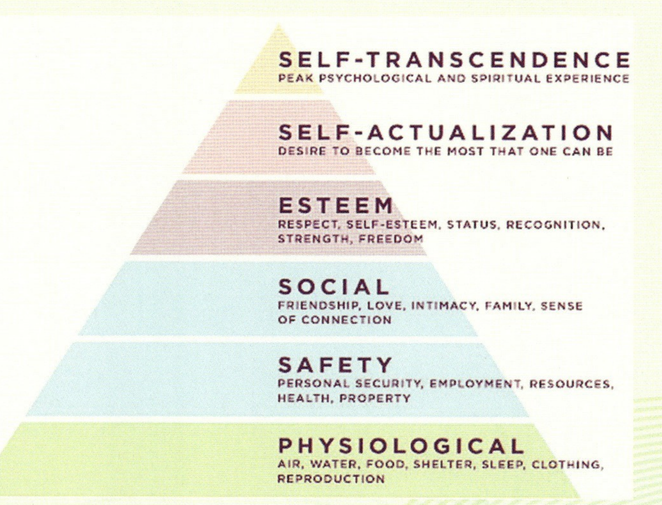

5.8. Maslow's revised pyramid.

However, Csikszentmihalyi took up the idea of Maslow's peak experiential states and interviewed thousands of people about the times in their lives when they had performed at their best. He began with experts—chefs, engineers, artisans, executives, surgeons, chess players, and dancers—and then threw his net wider to include Navajo sheep herders, Italian farmers, Chicago factory workers, Korean grandmas, Japanese teenage motorcycle gang members, and the most diverse variety of people he could find.

Every Decision and Action Leads Fluidly to the Next

The people he interviewed often used the term "flow" for these times of peak performance because every decision and action lead fluidly to the next. Csikszentmihalyi characterized flow as *a state of optimal experience where individuals are fully immersed and engaged in an activity.*[14]

He dissected the key components of flow. These include clear goals, immediate feedback, and the merging of action and awareness. He showed that flow leads to heightened performance and creativity, contributing to excellence in education, sports, business, and other fields.

Csikszentmihalyi examined the conditions that facilitate flow and the personal and environmental factors that spark it. He noticed that certain environments—like those that develop skills and contain tasks neither too easy nor too difficult—promote flow. He reviewed the impact of flow on wellbeing, suggesting that frequent experiences of flow lead to greater life satisfaction and happiness.

Finally, Csikszentmihalyi looked at the broader implications of flow, proposing that a society that fosters flow-inducing activities will enhance its members' overall quality of life. In Chapter 7, we'll see the exciting ways in which flow contributes to global human flourishing and performance.

The following story is a great example of how peak performers experience flow and the shifts it creates in their perspective. It also illustrates how closely the self-transcendent states of flow parallel the mystical experience.

FLOW PUSHES AN OLYMPIC RUNNER BEYOND HER LIMITS

Alexi Pappas is a runner, filmmaker, and author known for her dynamic achievements both on and off the track. She represented Greece in the 10,000 meter run at the 2016 Rio de Janeiro Olympics. She cowrote and starred in the feature film *Tracktown*. She is the author of the book *Bravey: Chasing Dreams, Befriending Pain, and Other Big Ideas.*[15]

Her life has been filled with both triumphs and tragedies. Her mother committed suicide when Alexi was just four years old. "It felt like something was taken from me…and it would have been really easy to make a rule of life that the world was a place of scarcity where things would continue to get taken from me," Pappas shared.[16]

But she chose to see her challenge in a positive light, observing, "When my mom passed away, I chose to believe that even though I could not have her, I could have everybody else; other mentors, female figures, and of course the lessons that I learned myself along the way."[17]

During her Olympic event, Alexi vividly recalls entering a state of flow that transcended the physical realm. As the starting gun fired, she surged forward, her body seamlessly synchronizing with the rhythm of the race. Each stride felt effortless, her breathing perfectly timed with her movements. Alexi felt as though she had tapped into a limitless reservoir of energy and focus.

5.9. Alexi Pappas.

As the race progressed, something extraordinary happened. Alexi's consciousness detached itself from her body and rose above the track. She could *see herself from above,* a focused figure in a sea of competitors.

This "out-of-body experience" (OBE) was empowering rather than disorienting. From this vantage point, she observed her every move with crystal clarity, noting how her legs propelled her forward and how her arms swung in perfect harmony with her stride.

In this state of deep flow, the noise of the crowd faded into a distant hum and time seemed to bend and stretch. Each second felt elongated, providing her ample space to make split-second decisions about her movements and tiny adjustments to her stride. She was both the runner and the observer, fully immersed in the present moment yet experiencing it from a plane of higher awareness.

This profound experience of flow allowed Alexi to push beyond her limits. She ran with a grace and efficiency she'd never known before. Crossing the finish line and setting a Greek national record, she felt an overwhelming sense of accomplishment, knowing she had accessed a rare and extraordinary state of being, one where mind, body, and spirit were perfectly aligned in flow.[18]

While stories like that of Alexi are drawn from the world of sports, where champions often describe entering flow, this state has been central to many artistic, creative, and scientific breakthroughs. Aboriginal artists enter the "dreamtime" in which they create symbolic art. The Renaissance sculptor Michelangelo "saw" the statue of the Pieta trapped in a block of marble. The ancient Greek mathematician Archimedes is said to have leaped from his bathtub and cried out "Eureka" as he discovered the law of displacement. Flow is a state that anyone can access.

5.10. Peruvian weaver near Machu Picchu creating traditional symbolic art in a delighted state of flow.

Biology—Not Psychology or Spirituality—Holds the Key

Biology is the key to flow. Biology is the way your body works and it's standard issue for all human beings.

Mind? Forget it. Not only is my mind radically different even from people as close to me as my wife, children, and grandchildren, it's even different from itself. My mind is so unpredictable that it can like something one moment and dislike it the next. It's full of confusion, conflicts, and garbage thoughts. Memories, fears, and preferences by the thousands swirl through it every day. When I step back and watch, it looks like a river filled with trash, swirling random scraps of rubbish here and there.

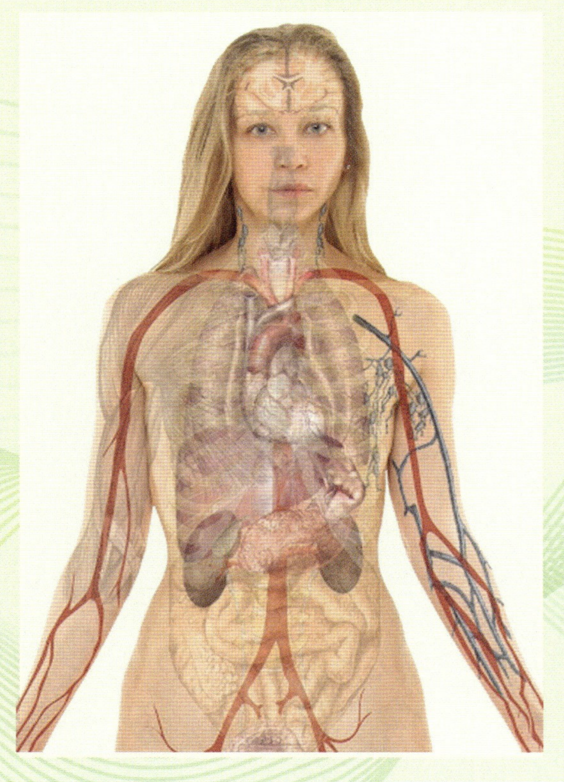

5.11. The basics of biology are consistent.

Can you relate?

Biology isn't like that. Your digestion works a certain way and it's consistent day after day. Ditto your heart, lungs, muscles, and bones.

Biology is reliable. You eat foods that you know from long experience work well for your body and you digest them easily. Swill a pint of double-chip ice cream or swig a quart of whiskey and you create havoc in your biology.

The results are predictable. Healthy inputs into your biology produce good results. Garbage inputs produce trashy results.

Cultivating SQ has far more to do with biology than with psychology or spirituality. Use the right biological inputs and SQ is enhanced. Forms of meditation that use biological signals to generate the Awakened Mind pattern in your brain are likely to be much more effective than *trying to use your unruly mind* to enter those blissful states of awakening.

The Religious Mishmash

Religions mix biology with theology. Eastern traditions tell you to meditate, Western traditions to worship, and they then give you helpful icons like saints or goddesses on which to focus your attention. They then surround the process of worship with elaborate belief systems recorded in voluminous scriptures.

So you follow their instructions. You sit quietly, breathe, become mindful, and relax your body—all in the midst of these religious trappings. You feel better and the priest says, "Thanks be to God! Glory to His name!"

Because we haven't understood biology, we think it's the theology that's making us feel better. We then invest in the religion, which often leads to disappointment. We get wound around the axle of all the hypocrisy, power plays, expectation, greed, hierarchy, and manipulation that goes along with organized religion. Many of us eventually walk away disillusioned.

That's where science steps in. It allows us to tease out biology from the religious window dressing.

For instance, in the Alzeimer's study in Chapter 2, patients who meditated got better and their brains regenerated. They increased gray matter in areas related to attention, executive function, memory, emotion regulation, and the inhibition of conditioned responses. They also lost gray matter in brain regions running the scripts of narcissism and unhappiness.

Priests and ministers would like us to credit their spiritual belief systems. If we mistook the true cause of healing for the mind, we might study the philosophy of Vedic scriptures, in this case Patanjali's *Yoga Sutras* from which the Alzheimer's meditations were drawn.

5.12. Sanskrit characters of the Yoga Sutras.

But when we look closely at what those patients were *really* doing, we see that they were *regulating their biology*. The first procedure they practiced was full-body relaxation, mindfully inhabiting each part of their body and releasing tension. They followed this with conscious breathing while stepping back from their thoughts and into the perspective of a witness.

This biological regulation is also key to EcoMeditation. The method is just a collection of science-based physiological cues, stacked one upon the other. It is not based on psychology or spirituality. It doesn't require you to believe in God, follow a guru, adhere to a faith tradition, or quiet your mind. It's based purely on biology.

The Essentials of Biological Regulation

While writing my first best-selling book, *The Genie in Your Genes,* I delved deep into the many techniques for meditation and stress reduction, like EFT tapping, neurofeedback, mindfulness, heart coherence, and hypnosis.

They all had passionate advocates, plus research showing they really worked. So I played around with combining them. I placed them in a simple sequence and found the sequence helped people meditate with ease. Here are the steps:

First, you *tap on EFT's acupressure points* to settle your body down and release physical stress.[19-20]

Second, you *relax your tongue on the floor of your mouth.* This sends a signal to your vagus nerve that it's time to relax deeply. The vagus connects with all your body's major systems: respiration, circulation, musculoskeletal, digestion, reproduction, immunity.[21]

Third, you n*otice the volume of space inside various parts of your body,* especially the spot between your eyes. This automatically generates big alpha waves, setting you up for the Awakened Mind brain-wave pattern.[22]

Fourth, you put yourself into heart coherence by slowing your breathing. You *breathe to the rhythm of six seconds per inbreath and six seconds per outbreath.* Though this breathing pattern is just a mechanical activity devoid of spiritual or psychological significance, it immediately regulates mood and emotion.[23]

Fifth, you picture your breath entering and leaving your body *through your heart* area, a visualization that has been found to engender an even deeper state of heart coherence.[24]

Sixth, you imagine *a beam of energy connecting your heart with a person who, to you, represents unconditional love.*[25] You then broaden this into *a feeling of compassion* and gratitude connecting you with every atom in the universe.[26] MRI scans of meditation masters shows that intense positive emotions like compassion produce big changes in brain function.[27]

Each of these practices done alone is effective. Combine them, and *you've changed your biology in a massive way*—no belief or spirituality required.

5.13. Engaging the energy of the heart and connecting to the infinite.

Reverse Engineering the Awakening Process

Teachers of spiritual practices like meditation and yoga discovered thousands of years ago that certain biological stimuli calmed the body and mind and facilitated the development of SQ.

For instance, over 2,000 years ago, Patanjali taught that the most reliable method of calming our turbulent minds is to regulate our breathing. Teachers therefore made breathing exercises—*pranayama* in Sanskrit—part of their instructions to students. The biological stimulus of slow breathing thus became entwined with the theology of Eastern religions. But breathing can be *separated from the tradition* and still produce the same calming physiological results.

Jamie Wheal, who has collaborated extensively with Steven Kotler in flow research, puts it this way: "Instead of spending decades trying to imitate wise old Tibetan monks, for example, never sure what was true mysticism versus mere mannerism, we could actually learn what makes them tick from the inside out." This includes "lower respiratory rates, more relaxed alpha-wave EEGs, and higher vagal nerve tone."[28]

You can take a regular person, discard all the trappings of religiosity, and have them mimic the physiology of a monk. Wheal points out that this is a "straightforward approach to human development that swaps out psychological rumination for physiological recalibration. It also spares us the endless and largely frustrating search for accidental peak experiences. ... We can tune the knobs and levers of our bodies and brains to trigger flashes of illumination" in a process of reverse engineering.[29]

When seeking flow states, Kotler advises against blindly following the advice of others who've done it before you. Someone else's path might work for you, or it might not. Human beings are very different. A set of beliefs and self-talk, for example, that work for one personality type might not work for another. But physiological cues like those cited here are biological and work for everyone.[30]

Kotler points out that *personality-based cues work only on people with similar personality types.* But biological cues, because they are fundamental to the ways our bodies function, work for almost everyone. Kotler observes: "Personality doesn't scale. Biology scales. It is the very thing designed by evolution to work for everyone."[31]

Where Does Religion Fit In?

The same might be said for religion: No religion scales up to a global size. Religions are a product of the times and cultures that produced them. But the *biological techniques that religions incorporate*—such as mindfulness, inner calm, moderation, and breathwork—scale up to work for every human being.

EcoMeditation is based on the concept that you can enter elevated states using a series of simple physiological cues like the breathing rhythm and tongue relaxation described previously.

This doesn't mean you'll never develop a spiritual practice or you'll discard all the philosophical discoveries of the masters. It means you won't mistake one for the other and *you can choose any spiritual tradition that speaks to you, or none at all.* Either way, you use EcoMeditation's biological cues to cultivate the transcendent experience of SQ.

5.14. The trinkets are just window dressing.

That doesn't mean religion plays no role in the process of awakening. Religions are repositories of ancient knowledge, the "wisdom of the tribe" curated and passed down from generation to generation. They're full of inspiration and insight. Who can read Paul's great paean to love (1 Cor. 13) and not be inspired? "If I speak in the tongues of men and of angels, but have not love, I am only a resounding gong or a clanging cymbal."

How can you not be moved by Jesus's Sermon on the Mount? "Blessed are the peacemakers… Love your enemies and pray for those who persecute you." Or from Isiah 40:31 in the Jewish scripture: "But they who wait for the Lord shall renew their strength; they shall mount up with wings like eagles; they shall run and not be weary; they shall walk and not faint." Or Buddha's admonition to "radiate boundless love toward the whole world" or, from the Bhagavad Gita, "Love spontaneously gives itself in endless gifts."

Paul advises us to "think on these things" and Patanjali's *Yoga Sutras* tell us the way to eradicate our habit of negative thinking is to deliberately substitute positive ideas like the ones above. We do this over and over again—for a thousand lifetimes if that's what it takes.

Belief Reinforces Biology

These are all helpful practices. They also engage *belief,* recruiting the placebo effect to *reinforce the biological stimuli*. Research has even shown that the greater the emotional fervor we bring to our spiritual practice, the more pronounced the brain changes that ensue.[32]

But biology is the foundation. Pair the wisdom of your favorite religion with breathing, mindfulness, heart coherence, and self-hypnosis, and you'll develop SQ much faster. In the study Judith Pennington published on her first measurement of EcoMeditation, she wrote: "EcoMeditation produced extraordinarily high levels of Gamma Synchrony…participants acquired elevated brain *states normally found only after years of meditation practice*. EcoMeditation facilitated participants' ability to induce and sustain the alpha brain waves characteristic of high-level emotional, mental, and spiritual integration."[33]

Using your biology gets you there much faster than your mind. If you reinforce the activation of your body's systems with your beliefs or religious faith, you enhance the effect, as this touching story by Erica Doolan shows.

THE ANGEL AT THE AIRPORT

By Erica Doolan

I'm so blessed that I was introduced to Youth for Christ (YFC) when I was in college. As a kid, I found church pretty boring. But the people in YFC were on fire and I learned to speak in tongues.

I also learned many other Christian skills like controlling your words and thoughts, tithing, and service to others. I've had a Christian life coach working with me every week for almost 20 years. I also used energy medicine, especially Reiki, to balance my body.

I discovered EcoMeditation a few years back, and it really helped me focus. I had a hard time meditating before that, but EcoMeditation made it easy. I felt so good that I decided to do it for 1,000 days in a row.

Angels began to show up in my life.

One time, after I had just landed, exhausted after a long flight from seeing my daughter, I discovered my suitcase was missing. All the other passengers had picked up their bags. The conveyor belt was empty. Frustration turned to panic.

Out of nowhere, this woman appeared beside me. She had a glow to her.

"You're looking for something," she said. I told her my luggage was lost, fighting back tears.

She smiled serenely and said, "Follow me." There was kinda this look in her eyes, and I trusted her. She just glided through the crowded terminal. We reached a dark corner, and she pointed to a lonely green suitcase, my suitcase, propped up against the wall.

"How did you...?" I turned to her and began asking.

But she had vanished into thin air. I stood there, holding my luggage, heart pounding with disbelief.

I've had lots of encounters with angels like that in the past few years. Now I have my own angel I can see and talk to any time I want.

You don't need to have faith like Erica's to reap the benefits of SQ. However, if your religious beliefs reinforce your meditative experience, you compound the biological benefits of spirituality.

The Brain-Wave Patterns of Peak Performers

In the 1970s, after Max Cade's retirement, Anna Wise expanded on his groundbreaking work. As well as mystics, she began to examine the brain-wave patterns of peak performers. Wise applied the Mind Mirror technology to people at the height of their creative and professional abilities. Her subjects included artists, executives, scientists, mathematicians, and inventors.

She meticulously analyzed the ratios of alpha, beta, gamma, theta, and delta waves in their brains, just the way Max had done with mystics. Wise was surprised to discover that their brains exhibited the same Awakened Mind pattern that Cade had identified in spiritual masters such as yogis, Zen masters, and faith healers.

But she didn't see the Awakened Mind pattern all the time. Just as Max saw the pattern peak when his subjects were in transcendent mystical states, Wise saw them peak when her subjects were in flow states. These were moments when they were performing tasks that required creativity, problem-solving, deep focus, and the other characteristics of flow.

Wise observed that, in peak performers, "adding more relaxed, imaginative alpha waves to the beta of ordinary thinking, and by also awakening the slower theta waves associated with deep meditation, creativity, and access to the subconscious," led to the enhanced creativity and problem-solving abilities inherent in flow.

In fact, in flow, the most executive part of the brain, the dorsolateral prefrontal cortex, actually deactivates.[34] The same happens in EcoMeditation.[35] This is counterintuitive—why would the sharpest part of the thinking brain *quiet down* in states of high performance?

One reason offered by Steven Kotler is that the brains of people in flow are at the border of alpha and theta. They may jump up to beta and do some cognitive processing when they have to make a decision. But if they're good at flow, they'll drop back to that theta-alpha border rapidly. This lowers their cognitive load and makes their brain work more efficiently.[36]

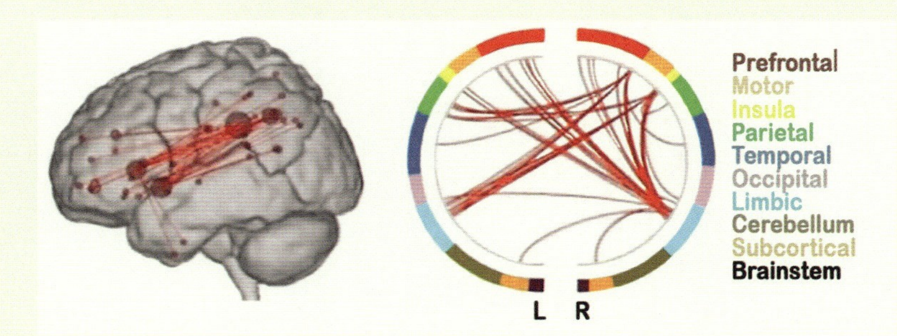

5.15. Connections between regions of the brain in a highly creative state.

Take a close look at the image above. It shows the results of a study examining the brain function of unusually creative people. The researchers found these people had a high degree of functional connectivity between all regions of their brain. Notice the list on the right, showing how many disparate and distant brain regions were involved. Look at how many lines connect them all. In extraordinary creatives, the whole brain works together as a *coordinated system.*

They weren't hammering away at problems using only the "left brain" top-down thinking of the dorsolateral prefrontal cortex. They were using all the brain's resources in a smooth, coordinated synchrony in order to meet their challenges. People in flow are "whole brain" thinkers.[37] They are activating many of the regions of the Enlightenment Network simultaneously.

Neuroliminals

That led me to the insight that the four circuits of the Enlightenment Network could be developed, one by one. Just like we work out at the gym and focus on growing a muscle group, like our chest, back, or legs, we could systematically focus on a single circuit. We could isolate that circuit and work out there intensively, sparking neural growth. Once developed, we could move on to the next circuit.

As a result, I made a series of recordings focused on activating a particular circuit. I call these Neuroliminals. They develop the four circuits we identified in Chapter 1, starting with the Emotion Regulation

Circuit. Having the ability to regulate negative emotions like anger, frustration, and resentment frees us from the petty irritation that constrains the happiness of so many people. When we aren't emotionally reactive anymore, horizons of possibilities open up for us. You'll find a downloadable Neuroliminal track in the Extended Play section of this chapter.

The Two Paths to the Awakened Mind

Anna Wise's research validated and extended Max's findings by demonstrating that the Awakened Mind pattern was not exclusive to spiritual practitioners but was also present in peak performers when they were in flow. That highly connected brain pattern is characteristic of both meditators and peak performers.

It became apparent there were two paths to the Awakened Mind brain-wave state—one secular, one spiritual. Peak performers, whether in the fields of science, business, music, athletics, art, or education, often describe their most creative moments as religious experiences.

It's not a figure of speech; peak performers in flow have mystic-like brain waves. Science has now demonstrated that the *flow states of peak performers are a second path to Bliss Brain.* Flow states can fire and wire the circuits of SQ as surely as mystical practice can.

Steven Kotler says that meditation and peak performance are the two primary ways to achieve flow. He says that while these two practices look very different from the outside, "neurobiologically, they look the same."[38] In the Extended Play resources at the end of this chapter, you'll find a link to Kotler's list of the preconditions for flow, plus the practices for cultivating it in our daily lives

Head in the Clouds

"He spends so much time with his head in the clouds that his feet don't touch the ground." When I was five years old, I heard those words spoken contemptuously of a man in our congregation who spent much time in prayer and contemplation.

People in our church generally believed that you were either spiritual or material. Saints abandoned the material world with all its pain and confusion and anesthetized themselves with piety.

The rest of us got down to business and figured out how to survive in the dog-eat-dog "real world." If your head was stuck up in the clouds, you weren't going to be much good when it came to the next barn-raising. And I witnessed several of the "holiest" people being out of touch with the realities of money, family, and body.

My friend John Gray, author of the best-selling book *Men Are from Mars, Women Are from Venus,* spent his 20s as secretary to Maharishi Mahesh Yogi. The Maharishi introduced Transcendental Meditation to the West in the 1950s, becoming a spiritual guru to many, including the Beatles and other celebrities.

From this exalted spiritual vantage point, John had the opportunity to get to know the many spiritual leaders who came to mingle with the Maharishi. John noticed that most of them were physically unhealthy. Also, there were all manner of financial and sexual shenanigans taking place in their communities. Their heads were doing fine up in the clouds, but their lives on the material plane were often a shambles.

Since I'm a researcher, I decided to design a study to investigate this question. I'd already performed several studies showing that EcoMeditation produced the Awakened Mind pattern and accelerated SQ. With the help of Andrew Newberg and Peta Stapleton, I'd validated the Transcendent Experiences Scale.

I'd identified the close relationships between the four types of Samadhi described in Patanjali's *Yoga Sutras* and Jeffery Martin's Four Locations. I'd offered courses on SQ and the Short Path to Oneness, and measured people breaking through to transcendence when they used the biologically based techniques in EcoMeditation.

So I certainly had a fine track record of helping people get their heads in the clouds.

"But what happened next?" I asked myself. What happens when these people finish the course, go back to work, rejoin their families, and have to function in the real world? Are they floating in disembodied bliss, out of touch with money, relationships, and their bodies? Are they neglecting their health like the medieval ascetics and John Gray's spiritual teachers?

So I designed a study to find out. Participants were people taking a basic EcoMeditation course called *A 21 Day Walk With Your Higher Power.* You'll find a link to the course in the Extended Play resources at the end of this chapter.

Participants spent three weeks doing EcoMeditation each day, as well as a series of exercises designed to connect them with their Higher Power and develop their SQ. Before and after the 21 days and again after six months, I gave them all the usual tests for conditions like anxiety, depression, and happiness.

5.16. Participants at SQ retreat.

But I added in assessments of workplace engagement, professional productivity, and personal productivity. I wanted to find out if they were more engaged with their jobs as they evolved their SQ, or less. Were they more productive at work and at home, or did their new preoccupation with the All That Is make them less attentive and effective in other domains of their lives? Did they get better or worse at solving problems? Did their cognitive ability go up or down?

The Awakened Mind in the Workplace

The results of the study were fascinating. As expected, we found an improvement in psychological conditions like anxiety (–27%), depression (–56%), trauma (–15.1%), and happiness (+21%) after the course. Flow increased by +20%. We also found increases in transcendent experiences (+23%) and nondual states (+16%). When we followed up with people six months later, we discovered they had sustained their improvements in both.[39] EcoMeditation again proved itself to be a fast and effective method for getting your head in the clouds.

Gaining an Extra Day Each Week

Then came the "feet on the ground" measures. While their work engagement scores were better after the course, the difference was not statistically significant.

What was significant was their productivity both at home and at work. It soared. In just 21 days, they became 20% more productive. That's like *getting an extra day in your work week.*

Not only did they show short-term gains in productivity, their six-month results showed an uptrend in which they were 26% more productive than they had been at the start of the course. That's getting an extra day-and-a-half in your week *every single week now and into the future.* Their feet were definitely and firmly on the ground.

A key study was performed by Yosi Amram, the author of *Spiritually Intelligent Leadership.*[40] Amram examined the contribution of EQ and SQ to the effectiveness of business leaders. Participants were 210 team members and 42 CEOs. Team members scored their CEO's leadership abilities and whether or not they contributed to a sense of community, job satisfaction, morale, and commitment to staying with the company. Amram found that SQ was significantly correlated with *leadership effectiveness.*[41]

Similarly, Turkish researchers examined whether the SQ of top managers affected the performance of their organizations. The investigators recruited 374 industrial managers in Istanbul and measured their levels of achievement over the previous three years. They found that SQ was associated with significantly increased financial performance.[42]

A similar effect was found in a sample of 300 employees drawn from Pakistan's 10 largest banks. It examined whether increased SQ had an impact on the financial performance of their institutions. The researchers found that high-SQ team members produced *a greater return on assets* (ROA), the measure by which banks measure financial success.[43] When we reflect on the brain circuits activated by SQ, like those for emotion regulation and focus, it makes sense that people with increased capabilities in those areas would be better money managers.

SQ is also associated with a number of other positive outcomes at work. It equates with increased job satisfaction.[44-45] It improves performance in the workplace.[46] SQ boosts both life satisfaction and resilience.[47-48]

A noteworthy study compared the business performance of 20 top-level Norwegian managers against 20 low-level managers. The two groups were matched for age, gender, education, and type of organization. Analysis of their EEG patterns found that the upper-level managers had a greater integration of brain function, better able to *unite perception, strategy, and behavior* into successful performance. Their brains also ran more efficiently. They also had more frequent peak experiences of "flow" than the low-level managers.[49]

Feet on the Ground

This echoes the findings of productivity studies of flow, the secular cousin of SQ. A decade-long research project performed by an international consulting company called McKinsey asked high-level executives to rate their ability to solve difficult strategic problems.[50] When in flow, the managers reported a *fivefold* increase in their performance. A study by the US Defense Advanced Research Projects Agency (DARPA) examined flow states in armed services members. It found that their ability to master new skills and solve complex problems *increased by 490%*.[51]

These studies demonstrate that "head in the clouds" and "feet on the ground" are not an either-or proposition. It's possible to be *both* spiritually intelligent *and* materially productive. The increase in flow states measured in my Higher Power study shows that you can carry the Awakened Mind pattern from the meditation cushion to the office chair. There, it translates into productivity, creativity, and problem-solving ability.

FLOW PRODUCES A FIVEFOLD INCREASE IN BUSINESS

Alicia Cornfeld is a tech entrepreneur specializing in innovative software solutions for small businesses. She founded her startup, TechFlow Solutions, with a vision to streamline business operations through cutting-edge technology. She recalls a transformative period in her career when she focused on making flow work for her.

She was struggling with the typical challenges of running a start-up—overwhelm, distractions, and inconsistent productivity.

"Discovering Steven Kotler's work on flow was a game-changer for me; a colleague suggested it, and it completely transformed how I approach my daily tasks," she recalls. She decided to incorporate these strategies into her daily routine.

Every morning, Alicia dedicated the first hour to a meticulous process designed to prime her mind for flow. She began with a period of intense information gathering, reading industry reports, and brainstorming ideas, allowing herself to feel the initial frustration that Kotler described. "Once I started my mornings in the flow state, I found myself effortlessly tackling complex issues like strategic planning and product innovation."

This was followed by a short break, during which she engaged in relaxing activities like walking or meditating to release the struggle and clear her mind.

As she returned to her desk, Alicia found herself slipping into a deep state of flow. In this state, distractions faded away and her focus became laser-sharp. Tasks that once seemed daunting now flowed effortlessly from one phase to the next.

Her productivity soared and she was able to tackle complex projects with newfound efficiency and creativity. She says, "Flow allowed me to dive deep into optimizing our business operations, which significantly improved our overall efficiency."

Over the next year, the results were astonishing. Alicia reports, "Our business's dollar volume skyrocketed from $2 million to $10 million after I embraced flow practices, proving its immense value in driving growth."

She credits her ability to consistently enter the flow state as the key driver of this growth. By mastering her mental state, she unlocked levels of productivity and innovation that propelled her startup to unprecedented success

Alicia's story is a powerful example of how harnessing the flow state can transform not just individual performance, but an entire business.

My observation is that when it comes to life fulfillment and success, meditation and flow are two sides of the same coin. I recommend people begin their mornings with meditation. This frames the day ahead with equanimity, compassion, and inner peace.

Morning meditation also sets up the day for flow. You enter the Awakened Mind brain-wave pattern for a while, which primes your brain for creativity, resilience, and productivity in the ensuing hours. Start the day in flow and you're more likely to maintain it throughout the day. A Harvard study found that entering flow upgrades the brain's creative performance for 24 hours afterward.[52]

Team Flow

I've experimented with team flow in a number of contexts. My rationale is that since EcoMeditation and biological practices produce increased productivity and wellbeing when practiced by individuals, they *should improve the performance of groups as well.*

For instance, every board meeting of the nonprofit I founded, the National Institute for Integrative Healthcare (NIIH.org), begins with a group flow exercise. It takes 10 to 15 minutes. A couple of decades of experience has shown that after aligning all our brains in the Awakened Mind pattern, we work together smoothly and creatively to solve problems and promote our healing initiatives. We've played a part in over 100 studies, making a major contribution to science. We've also treated over 20,000 veterans through our Veterans Stress Solution, applying this scientific knowledge to solving the major social problem of PTSD.

5.17. Team flow.

I've served on the boards of several nonprofits that did not use any preparatory rituals to induce flow and a positive mindset at the beginnings of meetings and instead just "got down to business." Despite the fact that most members were therapists, the meetings were often fractious and unpleasant, leaving people dispirited and exhausted at the end. Accomplishments were few and progress slow.

Many work teams report similar frustration with their meetings. A meta-analysis of more than a decade of research shows that 90% of employees feel meetings are "costly" and "unproductive." The data bear this out; the same study showed that employee productivity increased by over 70% when the number of meetings was reduced by 40%.[53] All this can be shifted *by generating team flow at the outset of every meeting*. I used this approach at a consulting project I undertook for a large family business.

SMITTY'S RETIREMENT

In the 1970s, a builder nicknamed "Smitty" founded a construction company in the Central Valley of California. It grew and prospered with Smitty as the patriarch and his children as employees. The company's board consisted of Smitty's wife, three sons, one daughter, and their spouses.

When he turned 75, Smitty retired and handed the reins to his second-oldest son, Len, since his eldest was more interested in hunting and fishing than continuing the family tradition.

Then the Great Recession hit. Demand for housing dried up. Smitty & Co. were left with hundreds of half-built homes, no buyers, vandalized properties, and huge construction loans to service. Len had to manage the relationship with the bank and panicked. He gave them faulty information and the bank was on the verge of seizing the whole business as collateral. Reluctantly, Smitty had to step up to the plate one more time and take over the reins of the business.

Their family board meetings often became blame games and degenerated into shouting matches. They'd hired an industrial psychologist to help, but his cognitive approach had proven fruitless against the biology of stress in which the family was enmeshed.

5.18. The Great Recession halted construction of thousands of homes.

I met with each family member several times individually and also began starting each group meeting with simple flow exercises like slow breathing. While some members of the family still needled others during the meetings, the emotional temperature of the whole group dropped considerably.

By the third meeting, the group energy had improved significantly. The family began to focus on solving immediate short-term problems instead of their usual habit of pointing fingers. By the fourth meeting, they were getting used to their new creative problem-solving mode and were able to make some painful but necessary changes to the business.

As I wrapped up my consulting engagement, I trained Smitty's daughter to lead the flow exercises at the start of each meeting. By this time, the culture of the family group had solidified into a positive new configuration. The downsized company was able to survive and eventually begin growing again. At long last, Smitty realized his dream of retiring.

Researcher Keith Sawyer has performed pioneering studies on group flow. Sawyer, a prominent psychologist and professor, was among the first to explore how teams can experience a collective state of optimal engagement and creativity.[54]

He discovered that when individuals collaborate in a highly synergistic environment, such as in jazz ensembles, improvisational theater, or innovative work teams, the experience can be significantly more fulfilling than solitary flow states. Sawyer's research suggests that group *flow can be three times as satisfying* as individual flow, as the shared challenge and mutual responsiveness within the group amplify the overall sense of achievement and enjoyment.

REWIRING THE CORPORATE BRAIN

Danah Zohar is a prominent physicist, philosopher, and management thought leader. She coined the term "spiritual intelligence" in 1997.

Zohar's work centers around the integration of SQ with personal and organizational development. She believes that SQ is the intelligence with which we address and solve problems of meaning and value, placing it at the heart of leadership and business excellence.

In her pioneering book *ReWiring the Corporate Brain,* Zohar applies the concept of SQ to the business world, arguing that companies need to go beyond traditional financial and emotional metrics to embrace meaning, purpose, fulfillment, and integration. She believes that businesses driven by SQ incubate innovation, ethical behavior, and sustainable long-term success.[55]

Zohar states, "Spiritual intelligence is the soul's intelligence. It is the intelligence with which we heal ourselves and others, and with which we make the world a better place." She emphasizes that SQ enables leaders to inspire and motivate others by connecting with their deeper values and purpose.

Zohar notes that, "Organizations that cultivate spiritual intelligence can better adapt to change, inspire their employees, and create a more meaningful and fulfilling workplace."

She asserts that businesses led by spiritually intelligent leaders are more resilient and better equipped to navigate the complexities of the modern world. Zohar's advocacy of SQ in the workplace showcases the potential of collaborative efforts to enhance creativity, productivity, and emotional wellbeing.

Imagine if flow states were induced not just in businesses, but in every other kind of group. There is ample evidence, for example, that mindfulness meditation helps children in school. Studies show that it improves both mood and test scores.[56]

Imagine children growing up with a tool to relieve test anxiety, social anxiety, existential anxiety, and climate change anxiety. The whole trajectory of their lives would be different and their chances of reaching their full potential infinitely greater. Research shows that spirituality "inoculates" children against depression and that developing SQ improves their mental health as well.[57] It also makes them more resilient.[58]

Analysis of prison programs shows that mindfulness alleviates anxiety and depression.[59] Imagine meditation and flow being used with groups in prisons, homeless shelters, juvenile detention centers, drug treatment centers, refugee camps, and other sites of acute suffering. It could make an enormous difference not just to the individuals involved, but to the whole of society.

Imagine if other group gatherings like education meetings, brainstorming sessions, charitable initiatives, product planning meetings, project reviews, and the many other ways in which people collaborate were all done in flow. Imagine the interpersonal friction being reduced as productivity and creativity soar. Imagine the increased efficiency, improved satisfaction, and better productivity that ensues.

5.19. Meditation and mindfulness in prison.

None of these settings are places we would normally associate with spirituality. If you're looking for enlightenment, you go to a temple rather than a prison, a retreat center rather than an elementary school. Yet right under our noses, as we're bringing flow and stress reduction techniques to these settings, we're facilitating mass enlightenment.

The application of flow to teams and groups is elevating our collective experience and accelerating our mutual awakening. We'll see in Chapter 6 how "enlightenment without spirituality" has been changing human brains for decades, but in ways so invisible they will surprise and delight you when you discover them.

Deepening Practices

The Deepening Practices for this chapter include:

- **Journaling Exercise 1:** When have you experienced the flow state in your life? Pick one or two experiences and record them in your journal. What were you engaged in at the time? A creative endeavor, performing, playing a sport, or simply being? Describe the experience.
- **Journaling Exercise 2:** If you grew up with a religious or spiritual tradition, consider the practices you were taught. Did they seem to relax your body and mind? Record the details in your journal. If you did not grow up with a religious or spiritual tradition, consider any messages you got about calming the body and mind. It could be as simple as the old practice of counting to 10! Record the messages and how or from whom they came to you.
- **Dawson Guided Neuroliminal:** Setting up Your Day for Flow

Get the meditation track at: SpiritualIntelligence.info/5.

Extended Play Resources

The Extended Play version of this chapter includes:

- A 21 Day Walk With Your Higher Power
- Thinkers50 video interview with Danah Zohar
- Podcast: Spiritually Intelligent Leadership: Yosi Amram and Dawson Church in Conversation
- Podcast: Flow: Steven Kotler and Dawson Church in Conversation
- Bonus section: Steven Kotler's Four Preconditions for Flow, plus Cultivating Flow in Daily Life

Get the extended play resources at: SpiritualIntelligence.info/5.

Updates to This Chapter

As new studies are published, this chapter is regularly updated. Get the most recent version at: SpiritualIntelligence.info/5.

Chapter 6

Compassion Drives Brain Evolution

Evolution Proceeds Very Slowly

The process of evolution through natural selection is incredibly slow, often taking millions of years to produce an entirely new species. Charles Darwin eloquently captured the essence of this gradual process in his revolutionary work *On the Origin of Species.*

In Darwin's words: "We see nothing of these slow changes in progress, until the hand of time has marked the long lapse of ages."[1] Evolutionary transformations are imperceptible over short timescales like that of a single human lifetime, with new species emerging only over vast geological periods. Or as Darwin put it, "no one can have watched the gradations by which each minute change has been effected" or could expect to see evolution progressing in his lifetime.[2]

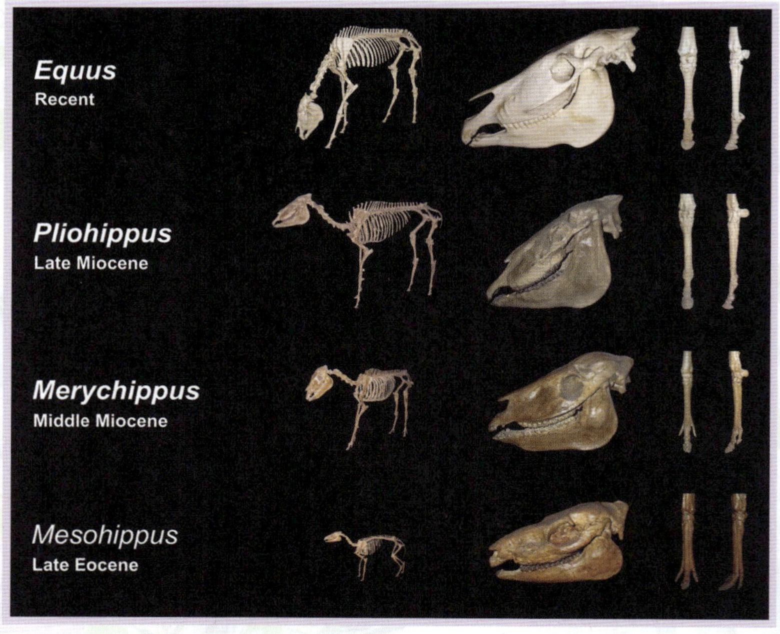

6.1. The fossil record shows changes to species like the horse over time.

Classic evolutionary theory holds that significant evolutionary changes occur through the accumulation of small, incremental changes over eons of time greater than thousands of human lifetimes.[3]

A case in point comes from a group of cave-dwelling crustaceans. These crabs and shrimp inhabit lightless underground caverns where vision is useless. Consequently, they have become blind, depending instead on smell and touch to find their way through the dark caves.

Researchers examining the brains of these subterranean creatures and comparing them to their surface-dwelling relatives discovered that, along with their loss of sight, *the brain regions responsible for vision are also shrinking*. Simultaneously, the brain areas governing touch and smell are expanding.[4] "It's a nice example of life conditions changing the neuroanatomy," remarked Martin Stegner, the study's lead author, in an interview with the BBC.

The brain changes in these crustaceans have unfolded *over the course of 200 million years.* While this may not seem rapid, *Popular Science* editor Rachel Feltman says that it's "a relatively short time, in the evolutionary scheme of things."[5] Even rapid evolutionary leaps, like the "Cambrian Explosion"—in which dozens of new life forms suddenly appeared in marine animals, leading to most of the body shapes we know today—took *12 million years.*

With Occasional Leaps

However, such gradual change doesn't always show up in the record of the fossils left behind by extinct lines of organisms. If evolution always took millions of years, we should find examples of all the intermediate stages between one species and the next.

For many creatures, we don't. There are gaps, with old species remaining unchanged for millions of years, then new species suddenly appearing. These anomalies perplexed paleontologists for decades.

Bryozoans are tiny marine animals under a millimeter in size. They clump together in lacy or fan-shaped colonies that look like moss. They first appear in the fossil record around 500 million years ago.

Fossils show that Bryozoans lived for millions of years without changing appreciably. Then, suddenly, for brief intervals, they changed rapidly. New bryozoan species appeared in the fossil record, without intermediate forms. They then settled down again for millions more years without perceptible change.

6.2. Bryozoan colony.

The way classical evolution explained these gaps was by saying the "missing links" existed; we just hadn't discovered those fossils yet.

But in 1972, paleontologists Stephen Jay Gould and Niles Eldredge came up with a new theory to explain this puzzle. They called it "punctuated equilibrium" and posited that *evolutionary change occurs in relatively short, intense bursts,* rather than gradually over millions of years.[6]

Gould and Eldredge proposed that the gaps in the fossil record might instead represent real instances of rapid evolutionary change. Based on their extensive analysis of Bryozoans, they argued that species remain relatively unchanged for extended periods of time but then change dramatically during short periods.

These periods range between *50,000 and 100,000 years.*[7] That's a blink of an eye in evolutionary terms. As Gould and Eldredge stated, "The history of life is more adequately described as a record of stability punctuated by episodes of relatively rapid change."[8] Their theory of punctuated equilibrium gave us a satisfying explanation for the gaps in the fossil record, while also suggesting that evolution can take relatively quick jumps.

Plants and Animals Evolving Rapidly

What if evolution is occurring even more rapidly than the 50,000 to 100,000 year span of punctuated equilibrium? As more recent data has become available, researchers have identified examples of extremely fast evolution in the plant kingdom.

Silverleaf Nightshade

Silverleaf nightshade *(Solanum elaeagnifolium)* is a pretty weed with a wide range. It has purple flowers, prickly spines, and poisonous berries. Its habitat ranges from Greece to Texas to South Africa. Farmers detest silverleaf nightshade because it infests their fields, crowding out valuable cash crops. Their usual response is to mow it down.

However, the plant has adapted rapidly. The more it's mowed, the faster it evolves. Its large central taproot goes down farther and farther—five feet down in the *first generation* of mowed plants. It grows more spikes on its stem to deter the caterpillars who eat the flowers. The flowers themselves become more toxic, leading to lowered pressure from predators.

6.3. Silverleaf nightshade (Solanum elaeagnifolium).

Frequent mowing has produced the opposite of what's intended and spurred the evolution of a "superweed." These enormous evolutionary changes occurred in *less than 10 years.*[9]

Evening Primrose

The evening primrose *(Oenothera biennis)* is another weed that farmers aim to control, usually using pesticides rather than mowing.

But in carefully controlled studies, researchers found that certain evening primrose plants exposed to herbicides developed genetic mutations conferring resistance, allowing them to survive and reproduce. Their less-hardy neighbors died off, allowing the herbicide-resistant specimens to fill those ecological niches.

It took *only five generations* for the evening primrose to develop resistance to commonly used herbicides. This rapid evolution was facilitated by the plant's high reproductive rate and genetic diversity, which allowed resistant individuals to quickly dominate the population. Genetic sequencing revealed specific mutations associated with herbicide resistance, providing a clear genetic basis for the rapid evolution observed in the species.[10]

Blue Columbine

Not all such changes occur because of direct human intervention like mowing and herbicide spraying. The wild sunflower *(Helianthus annuus)* has rapidly evolved to adapt to different soil types without any human stimulus. Within just a few generations, distinct populations have developed specific genetic traits that allow them to thrive in two completely different soil types: sand and clay.[11]

Similarly, the mustard plant *(Brassica rapa)* has recently learned to flower earlier when it faces drought conditions.[12] Some populations of Colorado blue columbine have suddenly changed a single gene that results in them having no flowers but more seeds, making them more successful at reproduction.[13]

The scientists involved are frequently stunned by these changes. Zachary Cabin of the University of California, lead author of the blue columbine study, describes the excitement his team felt on the flight back from Colorado where they'd been analyzing the flowers. "The first time we really realized the pattern was at the airport on the way home," Cabin recalled.

Cabin was reading off data as analyst Scott Hodges entered the data in the computer. "Scott could see the pattern developing, because he had all the data in front of him, and was getting more and more excited." They quickly realized they had "caught evolution in the act." Hodges exclaimed that "evolution can occur in a big jump if the right kind of gene is involved."[14]

These are just a few of the many studies showing rapid evolution in plants. Together they demonstrate that plant species can evolve much more quickly than Darwin ever thought possible. But what about animals?

Tuskless Elephants

A striking example of rapid animal evolution is found in the tuskless female elephants of Gorongosa National Park in Mozambique. This trait evolved during the Mozambican civil war (1977–1992), in which both sides financed their armies through ivory poaching. Some 90% of Gorongosa's elephants were killed during this period. Biologists hypothesized that having no tusks could be an evolutionary response, making elephants less appealing targets for hunters.

6.4. Tuskless elephant.

Investigators utilized a combination of historical video footage and contemporary surveys to investigate this hypothesis. They discovered that before the war, approximately 18% of Gorongosa's female elephants were tuskless. Just 30 years later, this figure had increased to 50%. Computer models indicated that this rapid evolutionary change was unlikely to be due to chance. The researchers were able to isolate the gene responsible for the changes.[15]

Wetland Frogs

Researchers from Rensselaer Polytechnic Institute (RPI) have identified a species of frog that has evolved over a very short period of time. The changes were prompted by the use of salt, which is used to remove ice from roads in winter in many northern countries.

The research team examined frogs that inhabited wetlands near a parking lot constructed 25 years earlier and regularly salted in winter. Unlike larger bodies of water, wetlands can't dissipate road salt quickly, and in this particular body of water, salt concentrations were nearly 100 times higher than in wetlands far from salted roads.

The research team collected eggs from nine groups of frogs from various areas and waited till they hatched into tadpoles. They then examined whether tadpoles from the more salt-polluted waters had evolved a higher level of salt tolerance. They found that tadpoles collected from the parking lot wetland could tolerate salt levels that killed off tadpoles from *all eight other areas* with lower levels of salt.

The researchers were astonished to discover that "over the course of *just 10 generations,* these wood frogs evolved a much higher salt tolerance."[16] That's a lot quicker than the 50,000 years required for punctuated equilibrium.

Green Versus Brown Lizards

Evolution does not always require a direct human stimulus to move into high gear. An instance comes from cute little lizards I've seen on my visits to forests in the US state of Florida. Known as Carolina anoles or green anoles, for millennia they've inhabited the lower branches and trunks of Florida's trees.

Then invasive brown anoles from other parts of the Caribbean began to encroach on their habitat. The brown lizards were larger and more aggressive. With limited resources and increased competition, the green lizards were forced to adapt. Within months of the Caribbean invasion, they moved high into the treetops.

In this new environment, the thinner and smoother branches necessitated physical adaptations. Over *just 15 years and around 20 generations*, the green lizards developed larger toe pads and stickier scales to better grip the smooth branches.[17]

"The degree and quickness with which they evolved was surprising," said Yoel Stuart, one of the authors of the study that identified the changes. To put this in perspective, he added, "If human height were evolving as fast as these lizards' toes, the height of an average American man would increase from about 5 foot 9 inches today to about 6 foot 4 inches within 20 generations."[18]

The Catfish of Albi

I have visited France many times to teach and to enjoy the wonderful company of dear French friends. There, I became aware of a bizarre example of rapid evolutionary change: the catfish of Albi.

Catfish are found in every continent except Antarctica. There are over 3,000 species of these bottom-feeders, distinguished by their "barbels" or filaments of tissue protruding from around their mouths. They can exceed nine feet or three meters in length and weigh several hundred pounds or kilograms, though most are of more modest proportions. They can live for over 60 years.

But the French catfish have developed an ability found in no others. Since their introduction into Albi's river in 1983, they've learned to hunt and eat birds. This behavior, first documented by researchers in Albi in the early 2000s, represents a significant departure from the diet of aquatic organisms typical of catfish.

The way they do this is fascinating. They creep up above the waterline toward flocks of unsuspecting pigeons, using their barbels to sense their prey. When they get close enough, they lunge forward and, with luck, catch a pigeon by its leg. They then drag the unlucky bird into the water for lunch. You can see a video of the hunt in the Extended Play Resources at the end of this chapter.

"This predatory behavior of catfish represents an exceptional case of dietary adaptation to new prey and exemplifies the plasticity of their feeding behavior," said Dr. Julien Cucherousset, lead author of the study that first described the behavior.[19] It's a vivid example of rapid behavioral change driven by environmental pressures and opportunities. Pigeons, attracted to the water for drinking and bathing, inadvertently became a potential food source for the catfish, which quickly adapted to this newly available resource.

6.5. Catfish hunting pigeon.

The birds seem completely unaware of the slow approach of the catfish, perhaps because nothing in their recent evolutionary history suggests a threat might come from the water. Even more remarkably, once a pigeon has disappeared beneath the waves in the jaws of a catfish, the other pigeons fill the gap in their ranks and completely ignore the absence of their late friend. They seem oblivious to the danger, though evolution might soon revise their collective behavior as quickly as it did for that of the catfish.

There are many other examples of animals evolving rapidly. Peppered moths in England evolved darker coloration during the Industrial Revolution to blend in with soot-covered trees. Owls in snowy regions of today's Scandinavia are changing their plumage to darker colors as global warming renders white winters shorter.

Salmon are spawning earlier. Bedbugs in New York City have become 250 times more resistant to pesticides than those in Florida. Mice have become immune to Warfarin, a poison often employed to counter rodent infestations. House sparrows in urban areas have evolved shorter wings to navigate better through city environments. Sea turtles who just 30 years ago perceived humans as their most dangerous predator now crawl up on crowded beaches to sleep.

Rapid evolution has been amply documented in plants. It continues as we move down the evolutionary ladder to the level of single-celled organisms. Antibiotic-resistant bacteria can evolve in *as little as 11 days.*[20] It also continues up the evolutionary ladder, to lizards, frogs, catfish, elephants, and many other animal species.

So what about humans? Have our brains and bodies been static for thousands of years, or does science show that we are evolving as rapidly as the other inhabitants of our precious planet?

Rapid Genetic Evolution in Humans

Genetic analysis has given us a powerful tool for evaluating rapid evolutionary changes in human populations. Groups of people can develop significant genetic adaptations within relatively short periods, driven by environmental pressures and lifestyle changes. Following are some examples demonstrating how humans can rapidly evolve traits that enhance survival and flourishing.

Tolerance for Low Oxygen Environments

In the highest regions of the Himalayan mountains, including peaks like Mount Everest, levels of life-giving oxygen are much lower. At these extreme altitudes, the oxygen concentration is only half what it is at sea level. This drastic reduction is due to the decreased atmospheric pressure, which makes it much more difficult for people to breathe and sustain physical exertion.

Because it is so hard to maintain human life up there, the Tibetan plateau was one of the last areas of the planet to be settled by human beings. But the ability of Tibetans to thrive in these low oxygen conditions isn't just because they're hardy. It's because when they colonized these regions, their genes quickly adapted.

A study compared indigenous Tibetans, who live at altitudes above 10,000 feet or 3,000 meters, with Han Chinese from Beijing. The latter are genetically similar but live at an elevation just 100 feet or 30 meters above sea level.

The researchers found that the Tibetans' blood was genetically predisposed to produce greater quantities of the protein hemoglobin, which transports oxygen to all the body's cells. Though the Han Chinese and high-plateau Tibetans split off only around 3,000 years ago, the genetic changes occurred quickly enough to allow the Tibetans to colonize one of the world's most inhospitable regions.[21]

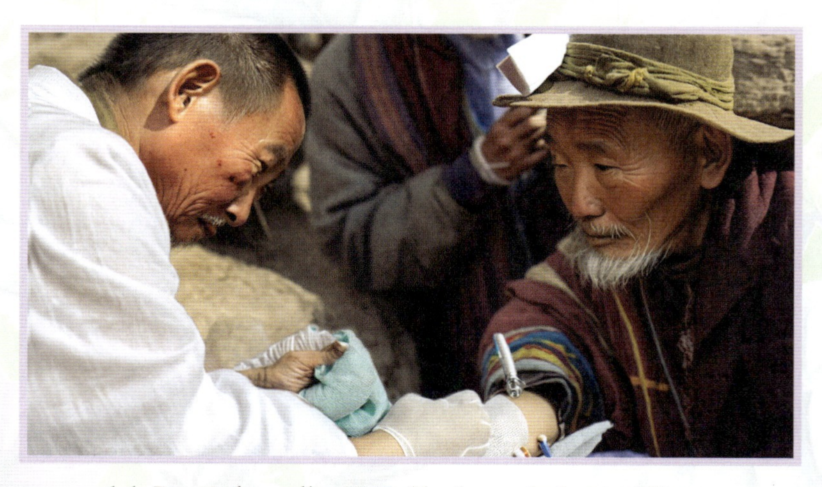

6.6. Researcher collecting a blood sample from a Tibetan DNA study participant.

The genome-wide comparison between the two groups was performed by evolutionary biologists at the University of California at

Berkeley. They identified 30 genes with DNA mutations (half of which are related to oxygen consumption) that became prevalent in Tibetans but not in the Chinese group.

One of the most notable adaptations is the increased frequency of the EPAS1 gene variant, which regulates the body's production of red blood cells. Another key mutation found in 90% of all Tibetans is found in only 10% of Chinese. The researchers found the *genomes of the two groups to be essentially identical except for the oxygen-related genes.*

Rasmus Nielsen, UC Berkeley professor of integrative biology, who led the statistical analysis, says, "This is the fastest genetic change ever observed in humans."[22]

Malarial Resistance

Evolution is highly invested in human beings staying alive long enough to produce children. That means helping our bodies defend themselves against deadly diseases.

One of the most-studied diseases against which the human body has recently developed new defenses is malaria. The ATP2B4 gene carries the body's code for building a major transporter of calcium in red blood cells, and variants of this gene harden the membranes wrapped around them. This produces a barrier that is difficult for the parasite to penetrate.[23] We evolved these gene variants between 2,800 and 7,500 years ago.

That's not the only way our bodies have recently evolved to fight malaria. Over a hundred different genes can cause a shortage of a protein needed to break down red blood cells. This makes it harder for the malaria parasite to infect them. Another recent mutation blocks malaria parasites from settling in the placenta, protecting unborn babies.

Besides malaria, evolution has also helped us develop defenses against leprosy, tuberculosis, and cholera. Certain genetic mutations make some people less likely to get these diseases. In populations where these illnesses were common, people with these protective genes were more likely to survive and have children, passing these helpful traits on to the next generation.[24]

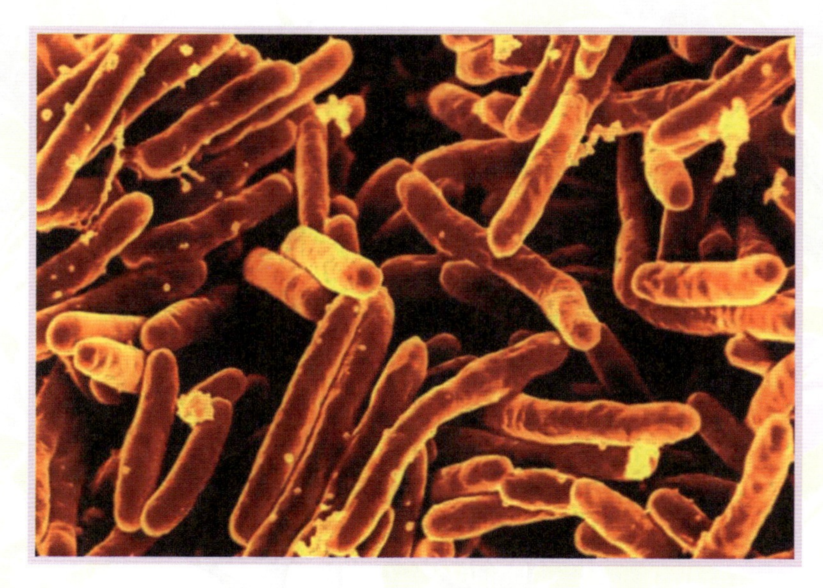

6.7. Tuberculosis bacteria.

Lactose Tolerance

The young of mammals subsist on their mother's milk for an extended period after birth. However, only humans can continue to digest milk after infancy. Mammals, including most humans, stop producing lactase (the enzyme that breaks down lactose, the sugar in milk) after weaning.

But some human beings continue to be able to digest milk into adulthood, a trait called "lactose tolerance." A mutation in the LCT gene, which enables the production of lactase, has become prevalent in certain populations. This mutation allows these adults to continue digesting milk. It's shared by about 25% of the human race while the remaining 75% of the global population remains lactose intolerant.[25]

A genetic mutation that emerged in Eastern Europe around 7,500 years ago enabled some humans to continue producing lactase into adulthood. Initially, they may have consumed cheese varieties like cheddar and feta, which have lower lactose levels compared to fresh milk, while hard cheeses like Parmesan have almost no lactose. Although these distinctions seem minor today, the capacity to digest high-calorie dairy products was a crucial survival advantage for our ancestors during Europe's harsh winters.[26]

Vanishing Wisdom Teeth

It's not just oral surgeons who are extracting third molars, better known as "wisdom teeth" from our mouths. Evolution is also lending a hand.

As human brains grew larger with each iteration of the *Homo* line, they pushed against the inner edges of our skull bones and narrowed our jaws. This made it hard for that third set of molars to emerge. We then discovered fire and began cooking our food, so we no longer needed those massive grinders to dispose of tough greens. Grains and starchy tubers didn't require the same determined chomping as the hunter-gatherer diet had demanded. This resulted in weaker jaw muscles and an increased risk of painful and deadly decay in impacted wisdom teeth.

A few thousand years ago, a genetic mutation emerged that prevented wisdom teeth from growing at all. Today, approximately one in four people are missing at least one wisdom tooth. This trait is particularly prevalent among the Inuit of northern Greenland, Canada, and Alaska, who are the most likely to lack at least one wisdom tooth.[27]

There are other examples of recent human evolution too, from the development of blue eyes to the ability to flush alcohol from the body. They show that rapid evolution through gene mutation isn't just found in plants and animals, but in humans as well. This biological fact leads us in surprising directions, as we'll see in Chapter 7.

Single-Generation Epigenetic Evolution in Humans

What's much faster still than evolution through gene mutation is epigenetic change. I became fascinated by epigenetics after the turn of the century. I wrote *The Genie in Your Genes* to explain the evidence showing that spiritual and emotional transformation triggers shifts in gene expression.

Epigenetic changes refer to modifications in gene expression that do not involve changes to the underlying DNA sequence. A gene can be turned on or off—expressed or silenced—without changing the structure of the DNA itself.

6.8. Like lights being dimmed up or down, genes can be expressed or silenced.

Think of the lights in your home. The circuits and bulbs are always there, just like the genes in your cells are always present. But whether the lights are on or off depends on whether you flick the switch. Epigenetic influences are what flips the switch.

Environmental factors such as diet, touch, and exposure to toxins can trip our genetic switches. So can ecstatic states like Bliss Brain, or stress states like Caveman Brain.

These changes can happen in moments. Get stressed and the CYP17 gene that codes for cortisol turns on in a second, as we saw in Chapter 1. Meditate and an epigenetic signal is sent to your cells telling them to silence CYP17 and express beneficial genes.[28] Meditation also sends epigenetic signals that reduce inflammation and rejuvenate your cells.[29]

These epigenetic effects are why levels of pleasure neurochemicals like oxytocin, anandamide, dopamine, and serotonin rise in the body during meditation, as I explain in depth in *Bliss Brain*. The *genes that code for those neurochemicals are being turned on epigenetically* by meditation.

One of the most fascinating aspects of epigenetics is that certain of these changes to gene expression can be passed down to our offspring. Traits can be transmitted within a single generation.

This process is much faster than even the fastest genetic mutation. In the studies described above, we see mutations occurring in periods ranging from 5 to 20 generations. Epigenetic changes can happen in a year or two and then be passed to the next generation. The following studies show the way epigenetic modifications are transferred to subsequent generations.

The Dutch Hunger Winter

At the end of WWII, Hitler retaliated against Dutch support for the Allies by cutting off food and other essential supplies to Holland. This led to a famine in which 20,000 people died. The lack of food triggered epigenetic changes in the Dutch population, and mothers who were pregnant had altered glucose metabolism and increased risk of cardiovascular diseases.

These changes were linked to modifications in the IGF2 gene, which plays a role in growth and development.[30] The molecular markers associated with these epigenetic changes were present not only in the children of these mothers, but also in subsequent generations. The stress experienced by the mothers had long-lasting effects on their descendants.

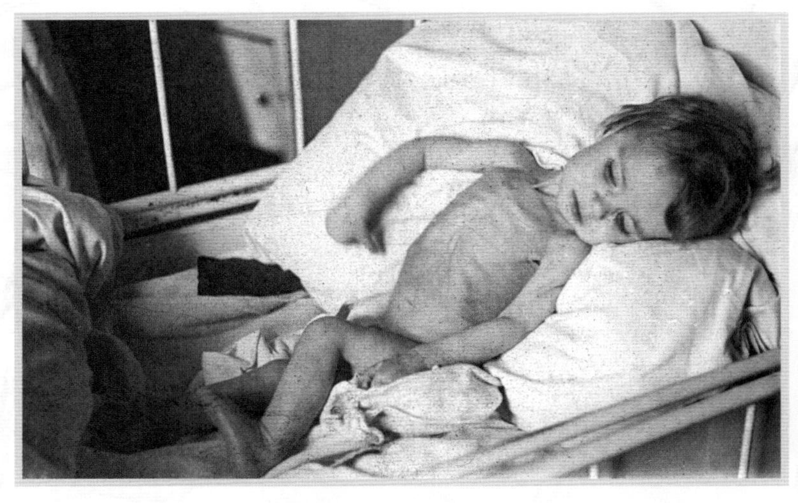

6.9. Starving child in the Hague, Holland, 1944.

A landmark study examined the impact of prenatal maternal stress on epigenetic programming in newborns.[31] The researchers found that infants born to mothers who experienced high levels of stress during pregnancy had altered expression of a gene called NR3C1, which is crucial for regulating stress hormones. These epigenetic changes were linked to cortisol spikes in the infants, indicating that the prenatal environment of a stressed mother's womb had a direct impact on the stress regulation mechanisms of the next generation.

Stress isn't passed to offspring epigenetically through mothers alone. Fathers can propagate stress through their sperm. When researchers stressed one of two groups of mice by spiking their cortisol levels to produce chronic stress, it changed epigenetic tags on their sperm. The researchers were astonished at the scope of the changes. They affected 2,382 different types of RNA. The offspring of those mice also exhibited behavioral changes that mimicked anxiety and depression in humans.[32]

The Effects of Smoking Extend from Mothers to Babies

Smoking is an epigenetic influence that damages cells. If mothers smoke during pregnancy, these epigenetic effects extend to their children. Among the switches tripped in babies born to smoker mothers are genes involved in lung function and the immune response.[33] Prenatal exposure to smoking leads to lasting epigenetic modifications that may influence the health of more than one subsequent generation.

The Lasting Impact of the Holocaust

The transgenerational effects of trauma were measured by examining the genomes of Holocaust survivors and their offspring.[34] The researchers found that both survivors and their children had altered expression of the FKBP5 gene, which is involved in the regulation of cortisol and the stress hormone system. These epigenetic changes were linked to increases in stress-related medical diagnoses. Not only did the daughters of Holocaust survivors show these genetic changes, *their granddaughters did too.* This study showed that *epigenetic changes are heritable and can be passed along through several generations.*

There are many other studies showing that epigenetic changes happen quickly and can be passed through multiple generations of human beings. Studies using animals have measured the effects extending through seven generations. This process is much faster than genetic mutation and can have profound effects on individual health and behavior. It can also change whole societies, as we'll see later in this chapter as well as in Chapter 7.

Rewiring the Fearful Brain

I've been part of the healing and wellness professional community for half a century. I've collaborated with, trained, or interacted with thousands of therapists and coaches.

One of the dark secrets of the world of professional therapy is that clients rarely change. Therapists and coaches desperately want to help them. They have all the skills, tools, and techniques required for transformation.

On the other side of the relationship are the clients. Clients desperately want to escape the behavioral, psychological, and spiritual patterns that keep them suffering.

Yet despite all this effort and attention, it's incredibly difficult for people to change ingrained behaviors. Caveman Brain kept our ancestors safe for hundreds of thousands of years and isn't willing or able to relinquish its protective role without resistance. Here are some examples.[35]

Change is Difficult and Rare

Studies show that *over half* of patients don't comply with instructions from their doctors. When glaucoma patients received eye drops to relieve the condition, fewer than half actually used them—even though the penalty for noncompliance was the *risk of going blind*. Only 19% of those making New Year's resolutions to modify a problem behavior maintained the change two years later. People getting psychotherapy for conditions causing them serious distress dropped out by, on average, the third session. When therapists confronted problem drinkers about their drinking, they began binging more.

JAIL ON CHRISTMAS EVE

For many years, before I met my wife, I shared a house with a dear friend named Jeff Anderson. A former alcoholic, Jeff had turned his life around after a car accident in which he broke his back. He was now a minister and also the leader of a large 12-step group.

I couldn't truly understand Jeff's story about how, in his old life, he'd continued drinking till he lost everything. So I went to some of the Alcoholics Anonymous meetings he facilitated. There I met Gerry, whose story astounded me.

Gerry had been consumed by fear for as long as he remembered. Anxieties about his career, relationships, and self-worth gnawed at him incessantly. To cope, he turned to alcohol. What started as a nightly drink soon spiraled into a debilitating habit of binge drinking. Every evening, Gerry found solace in the numbing embrace of alcohol, temporarily quieting the storm inside his mind. Drinking crept into his day, even his morning.

Recognizing that he was in a downward spiral, Gerry sought help through psychotherapy and coaching. His therapists delved into his past traumas and his coaches encouraged him to set goals and envision a brighter future. However, the grip of fear remained unyielding. Sessions often ended with Gerry feeling more defeated than before, his fears and failures only amplifying his sense of inadequacy.

Gerry went from disaster to disaster. After several warnings from his boss, he lost his job due to frequent absences and poor performance. His relationships crumbled, with friends and family distancing themselves from his volatile behavior. Money woes soon followed, leading to a mountain of debt and the eventual loss of his home.

No matter the problem, Gerry chose the same solution: another drink.

One Christmas Eve, Gerry was arrested for drunk driving. Sitting in the jail cell, he realized he'd reached the end of the road. The

6.10. The end of Gerry's road.

judge offered him the option of AA or jail, and Gerry joined Jeff's group. When Gerry shared his story with me, he'd been sober for nine years and had turned his life around completely.

What amazed me about Gerry's story was that he'd crashed through so many stop signs on his way to the bottom. Losing his job wasn't enough to motivate him to change his behavior. Nor was losing his friends, nor was losing his house. Gerry maintained his addiction despite the huge stack of negative consequences piling up.

When a behavior is based in fear, it's especially hard to change. Patterns that are crucial to survival, or that our brain has been trained to imagine are crucial, become hard-wired, even if they're as dysfunctional as addiction. They occur reflexively, below the level of conscious choice.

Prey animals, for instance, have evolved fixed neural circuits that help them escape safely from threats. In the same way, the circuits of Caveman Brain can light up in humans feeling threatened by life itself. When the fear response of Caveman Brain is active, our cognitive ability drops. But when the Enlightenment Network regulates our emotional responses, we become smarter.[36]

Addicted to Cortisol

While Gerry may have used alcohol as his drug of choice, I find many people to be addicted to a different drug that can be just as addictive: stress. Cortisol is the hormone of stress and most modern humans have elevated levels. I've tested people lounging in armchairs in luxury hotels and found cortisol levels more appropriate to running from *T. Rex* in Jurassic Park. Stressed people are literally addicted to their own high cortisol levels, making it hard to change.

6.11. Modern humans with easy lives can have cortisol levels as high as people facing deadly threats.

I also witness friends and family members failing to change. Extremely smart people I know personally persist with behaviors that have a 100% probability of resulting in disaster. Some of them are in the last part of their lives, yet decades of repeat failure and loss have been unable to shift them into new patterns. It's acutely painful to watch them setting themselves up for a fall yet one more time.

That's why I've been keenly interested in any approach that is more likely to disrupt the patterns ingrained in Caveman Brain. That's my motivation to teach EFT tapping, meditation, and body-based techniques because these are some of the few methods with extensive research showing they're able to produce rapid and reliable change.

There are now over 200 studies of EFT tapping, several on EcoMeditation, and thousands on other forms of meditation. They show that these methods are able to heal pain, anxiety, depression, PTSD, and autoimmune conditions. And intriguing brain studies have now shown that it might be possible to *modify the hard-wired fear networks of Caveman Brain* that keep us stuck in our predictable old patterns of suffering.

Extinguishing Fear in Mice...

In one study, researchers presented mice with a dark expanding disk overhead that looked like a predator moving towards them from above. The mice always fled the disk, even after it had appeared many times without jeopardizing the mice's survival. This indicated a threat response hardwired into their brains.[37]

The researchers then blocked access to their escape route. They also adjusted the contrast so that the disk could be lighter (less threatening) or darker (more threatening).

The mice eventually learned to ignore the disk, even though they remained responsive to other threats such as loud noises. "This suggests that escape is not simply reflexive but dependent on threat memory and is therefore under cognitive control," said Troy Margrie of the Sainsbury Wellcome Centre, corresponding author for the study.[38]

6.12. Mice can learn to extinguish even genetically encoded fears.

The study demonstrated that mammals can learn to overcome even strongly encoded fear responses. That's good news for us all, since it shows that, under the right conditions, *mammals can modify even basic neural networks that have been hardwired into their species for thousands of years.*

The researchers also tested mice living in social groups of 20 individuals against mice living alone. The isolated mice were much more likely to remain vigilant and reactive than the socialized mice. This echoes the research on meditators, which finds faster development of the Enlightenment Network in people who meditate together than in solitary hermits.[39]

...And Men

Remember that inspiring piece of neural wiring from Chapter 1, the ventromedial prefrontal cortex? It's the slice of tissue connecting the brain's executive centers with the limbic system. Part of the Emotion Regulation Circuit, it's the conduit activated by Tibetan monks and master meditators to calm the emotions. In people with major depressive disorder, it frays and signals start running backward. This enables the emotional brain ("I feel bad") to control the cognitive centers ("Think up a reason to justify me feeling bad").

An innovative human study examined whether stimulating the ventromedial prefrontal cortex with electrical impulses could extinguish fear. On the first day of the study, participants in two groups were exposed to a series of images. One of these images was accompanied by a mild electric shock administered by the researchers.

Then, in the experimental group only, the images were presented a second time, but on this round without shocks. The intention was to form a "fear extinction memory" (a memory of the fear-inducing stimulus) but this time *associated with no adverse effects.* The control group did not receive this fear extinction training.

Two days later, both groups were exposed to the images again. The experimental group received electrical stimulation of the ventromedial prefrontal cortex while the control group did not. Members of the control group still retained fear associated with the image. But the startle response and physiological stress did not return in members of the experimental group, indicating that their fear had indeed been extinguished.[40]

Christoph Szeska of the University of Potsdam, lead author of the study, summed it up by saying, "Electrical stimulation of the ventromedial prefrontal cortex—a critical relay that mediates the consolidation and recall of fear extinction memory—can block such return of fear."[41] That's the very circuit that meditation masters use to regulate negative emotion.

These and similar studies show that *it's possible to extinguish fearful memories* and escape the influence they produce on our behavior. When we develop the neural circuits that regulate the emotions that disrupt our inner peace, we gain the freedom to choose the thoughts, actions and behaviors that will serve our long-term good and that of the planet.

Jeffery Martin, describing the changes that occur when Seekers of awakening become Finders, puts it this way: "The average person lives with an experience of the world that is rooted in fear, worry, anxiety and scarcity…a feeling that haunts us in the background." But after awakening, "your discontent and fear is replaced by a sense that everything is fundamentally okay, that you are safe, whole, and fine just as you are." Fundamental wellbeing takes root and anxiety, stress, and depression evaporate.[42]

That's what it's like when the ventromedial prefrontal cortex, the emotion control knob of the Enlightenment Network, is active. Regulating the fearful brain frees us from dominance by our emotions. This allows us to make rational choices, opening the doors to genuine and lasting change.

Imagine a society in which everyone had access to fear-extinguishing techniques and used them to release their stress and lower their cortisol. We'd have a lot of happy people making wise choices. That's the promise of turning on the Enlightenment Network and cultivating SQ.

We're now going to jump from the individual to the social perspective, to see how SQ and the emotion-regulation functions of the Enlightenment Network can produce cultural change on a massive scale.

Cumulative Culture

Today's culture and technology are the result of thousands of years of accumulated and remixed cultural knowledge. This is "cumulative culture"—the gradual accumulation of insights and improvements over generations.

Cumulative culture has profoundly influenced human evolution. It has allowed humans to adapt to a variety of environments and overcome enormous challenges. Think of the first person to use fire passing on the knowledge to all the others in the tribe. Subsequent generations learn to use fire not just to stay warm but to cook food. Later generations, to clear terrain and harden tools. But when did our ancestors start building on each other's knowledge, setting us apart from other primates?

6.13. Early humans exchanged knowledge through cumulative culture.

To determine the answer to this question, Charles Perreault and Jonathan Paige from Arizona State University analyzed changes in stone tool manufacturing techniques over the last 3.3 million years. They found that around 600,000 years ago, humans began rapidly accumulating technological knowledge through social learning.[43]

Perreault said our species "has been successful at adapting to ecological conditions—from tropical forests to arctic tundra—that require different kinds of problems to be solved."[44] This success is largely due to cumulative culture, which enables human populations to build on and recombine the solutions of prior generations, developing new and complex solutions quickly.

To track the progression of cumulative culture, the researchers compared the complexity of ancient stone tools to those used by nonhuman primates and by inexperienced people attempting to create flint blades.

THE OBSIDIAN ARROWHEAD

I've seen countless stone age arrowheads in museums and gift shops. This technology shows up in Paleolithic sites all over the world, discovered by early humans in many geographically unconnected societies around the same time.

When you look at such an arrowhead, you might dismiss this technology as simple and primitive. What's so special about chipping pieces off a rock to create a sharp edge?

If you'd like to try a fun personal experiment, try creating one yourself. This arcane art is called "knapping"—the process of shaping stone by striking it.

6.14. Knapping an obsidian arrowhead.

I tried this at a quarry in Clearlake, California, in which early Indigenous people mined obsidian, a black glass volcanic rock. It was a humbling experience. Though I'm good with tools and motor skills, I found that striking the rock in just the right place demanded all my attention. If you strike the surface just a millimeter away from your target, a big piece flakes off and the fine edge of the instrument is lost. You have to start all over again with a fresh rock.

Knapping is like threading a needle. It requires a degree of visual focus like few tasks modern humans do. It has to be sustained for an extended period of time. You can spend an hour hammering perfect flakes out of the surface, but if your brain gets fatigued and your attention wanders near the end of the process, you can ruin the whole piece in an instant.

Trying to create my own arrowhead gave me a new respect for our "primitive" forebears and the cumulative culture that enabled them to master vital skills like knapping a piece of flint.

In their analysis, Perreault and Paige counted the number of steps in the tool-making process, as revealed by the tools left behind. That enabled them to assess the complexity of the procedure. They found that from 3.3 million to 600,000 years ago, stone tool manufacturing remained simple, with sequences involving just one to seven steps. But around 600,000 years ago, there was a rapid increase in complexity, with up to 18 steps used to create sophisticated blades. This suggests that cumulative culture began to take hold during this period.

Tools, in turn, led to significant changes in brain size, lifespan, social organization, and biology. The dawn of complex tool-making was accompanied by the controlled use of fire, hearths, and domestic spaces —all essential components of cumulative culture. Later humans learned to combine a stone blade with a wood handle. This leverage represented a major technological advance.

Jonathan Paige notes that, "By 600,000 years ago or so, hominin populations started relying on unusually complex technologies, and we only see rapid increases in complexity after that time as well. Both of those findings match what we expect to see among hominins who rely on cumulative culture."[45]

Science is one of those human endeavors in which cumulative culture is readily evident. Today's scientists "stand on the shoulders of giants." We read each other's work and build on others' insights. As the amount of available information increases, we have an ever-richer store of material with which to create new discoveries.

Spirituality is a second domain in which cumulative culture accelerates progress. We can read the stories of people who've had transcendent experiences before us. Great scriptures can inspire us with timeless truths. We can seek the company of like-minded souls; Buddhism considers the community of the faithful—the *sangha*—one of the three pillars of practice. We can meditate in groups, which produces faster epigenetic shift and neural rewiring than meditating solo.

We can share the nature of our peak experiences with other seekers. We can study with master teachers from the great wisdom traditions, and learn the steps of the Long Path and the Short Path to oneness. In all these ways, SQ is spread through cumulative culture.

CUMULATIVE CULTURE IN EARLY SPIRITUAL TRADITIONS

One of the primary ways in which Buddhism fosters enlightenment is community in the form of the *sangha*. A traditional Buddhist sangha consists of four main groups: monks, nuns, laymen, and laywomen. The sangha is one of three pillars of practice, the others being the Buddha and the *dharma* (teachings).

The first sangha emerged from disciples who abandoned worldly life to accompany the Buddha and learn from his teachings. After the Buddha's death, his followers continued living collectively, traveling and subsisting on charity. Every two weeks, on the days of the full and new moon, these first Buddhists congregated to reaffirm their communal bonds and recite key beliefs like the Eight Noble Truths.

Cumulative culture gradually evolved this practice into a settled form of community life. Buddhists began to spend the rainy season in contemplative retreat, inspired by the great teachers who followed Gautama. Today's sangha adheres to the disciplinary rules called *vinaya* preserved in the sacred texts and draws inspiration from modern teachers of the various Buddhist lineages.

Similarly, in the early days of Christianity, community played a vital role in the lives of Jesus's first disciples. These men and women were drawn together by their shared experiences and profound belief in Jesus as the Messiah. They formed a close-knit group, supporting each other through trials and triumphs.

6.15. The early Christians formed tightly knit congregations for mutual sharing and support.

After Jesus's crucifixion and resurrection, his disciples faced uncertainty and persecution. They gathered regularly, not only to pray and worship, but also to share meals and resources. This communal life strengthened their faith and provided them with a sense of unity and purpose.

The disciples were known for their generosity and compassion, caring for each other and for those in need within their community. They pooled their resources, ensuring that no one among them was in want. This solidarity was central to their identity as followers of Jesus, reflecting his teachings of love and service.

Cumulative culture produced a growing sense of community as Christianity spread. Jesus's early followers preached his message of love and forgiveness. They welcomed new believers into their fold, establishing churches in various cities and regions. Despite facing persecution and challenges, their commitment to each other and to their faith remained steadfast.

The early Christian community's bond was not only spiritual but also practical, as they navigated the complexities of living out their beliefs in a world that often misunderstood or opposed them. Through their shared experiences, trials, and joys, they forged a community that would become the foundation of the Christian faith, inspiring generations to come with their example of consistency, love, and unity.

Culture Accelerates Human Evolution

After thoroughly reviewing the existing research and evidence on human evolution, scientists Tim Waring and Zach Wood have identified a "special evolutionary transition" in which culture is accelerating the development of human society. They assert that culture, including learned knowledge, practices, and skills, is today driving evolution even faster than natural selection.[46]

Cumulative culture changes society rapidly because cultural influences spread quickly and are constantly updated. Even epigenetic shifts are confined to offspring in the same family, while cultural transmission is highly flexible and can incorporate knowledge from peers and experts beyond the parental scope. The effect is magnified in today's wired world.

Waring and Wood show that the synergy of cultural and genetic factors has driven crucial human adaptations, such as reduced aggression, enhanced cooperation, collaborative skills, and social learning capabilities.

Since groups are the primary vehicles of culture, Waring and Wood conclude that *evolution itself has shifted to be more group-centric.* Waring suggests that "Humans are evolving from individual genetic organisms to cultural groups which function as superorganisms, similar to ant colonies and beehives."[47]

What does history tell us about the contribution of genetics, epigenetics, and cumulative culture to the development of SQ? When we look back at the trajectory of recent human evolution, it shows surprising trends. It also points to a future far more peaceful and compassionate than today's headlines would suggest, as we'll see in Chapter 7.

Cognitive Biases That Obscure the Big Picture

Reading the news of the day is like looking at the stitches in a bed quilt. You're zoomed in to the details. You have an extremely clear idea about what a single cubic millimeter of the quilt looks like. The mistake most of us make is to extrapolate our knowledge of that tiny fragment to the whole. We use our fragmentary picture to make general assumptions about the quilt. This usually leads to a completely inaccurate assessment.

Human beings have a number of such biases. The *recency bias* gives more weight to recent information than to older data. Recent experiences are vivid in our memories, crowding out earlier information that might have given us a different perspective.

Another common cognitive error is the *confirmation bias.* We tend to seek out and give preference to information that confirms our preexisting beliefs while disregarding or minimizing evidence that contradicts those beliefs. Social media and conversations with our friends often become echo chambers that reinforce our misconceptions.

Anchoring bias occurs when people rely too heavily on the first piece of information they receive—the "anchor"—when making decisions.

You see a red stitch and you assume the whole quilt is red. This initial information influences your subsequent judgments though it may be irrelevant or misleading.

6.16. Biases are inherent and often unconscious.

The *availability heuristic* skews our perception toward overestimating the importance of information that is most readily available to us, such as vivid or recent memories or facts that are easily retrieved. We also perceive *dramatic or emotionally charged events* as more common than they actually are.

These cognitive biases lead to skewed perceptions of reality. They are all at work when we read the news. Today's upsetting headlines are like the individual stitches, and when we're enmeshed with our own story, we miss the big picture.

Perspective Taking

Neuroscientists study a function of the brain and mind called "perspective taking." It involves the ability to step out of entanglement with one's own biased thinking and notice the bigger picture. This enables us to perceive other people's viewpoints, as well as the context of the wider world.

This cognitive skill engages brain regions such as the medial prefrontal cortex, temporoparietal junction, precuneus, and other structures of the Enlightenment Network. Activation of these brain areas facilitates empathy and social cognition, shifting our mental state towards greater understanding and objectivity.[48]

Perspective taking enables us to zoom out and look at the whole quilt. This shows us a completely different picture of reality. It demonstrates that human society has been evolving rapidly for several hundred years and that progress has speeded up since the turn of the century.

500 Years of Increasing Compassion

Hard data on the direction of society and consciousness for the past 500 years show that human beings are evolving to be more compassionate. While the news headlines show us a red thread, the long-term scientific picture is of a colorful quilt.

Go back over 500 years, and while there are inspiring examples of compassion like Buddha, Jesus, and Lao Tse, they were outliers. But in the past 500 years, the whole of human civilization—involving billions of individuals—has *evolved rapidly to become more compassionate.* The following are several striking examples of this trend and summaries of many others.

Child Labor

For thousands of years, child labor was an accepted norm across the globe. There was no social debate or controversy surrounding the practice.

In the coal mines of Britain during the 1800s, children as young as eight served as beasts of burden, pushing heavy tubs of coal through narrow underground tunnels. These young laborers, sweating profusely and inhaling hazardous black coal dust, would emerge from the mines covered in grime and soot, their bodies prematurely aged and weakened. Many died young of "black lung disease."

In the early 1900s, children in the United States toiled in textile mills for up to 12 hours a day, six days a week. They often worked in dangerous conditions, operating hazardous machinery without safety measures, sometimes suffering severe injuries.

A noisy corner.—Rivetting in the boot shop.

6.17. Child labor was common even at the start of the previous century.

In India, children have historically been employed in industries such as carpet weaving, where they worked long hours in cramped and poorly ventilated conditions, developing eye strain and respiratory illnesses.

The movement to abolish child labor gained momentum through significant historical events and reforms. One pivotal moment was Britain's passage of the Factory Act of 1833, which regulated how many hours a week children could work and required factory inspections to enforce these limitations.

Another landmark event was the1916 Keating-Owen Child Labor Act in the United States. It prohibited the sale of goods produced by factories that employed underage children. Although initially struck down by the Supreme Court, it laid the groundwork for future reforms.

The tireless advocacy of reformers and the rise of compulsory education laws led to the abolition of child labor in most parts of the world in the early 20th century. While *for thousands of years human beings accepted child labor without question,* we changed our collective minds 180 degrees in just a century.

WOMEN'S RIGHTS

In the mid 1800s, the term "women's rights" had never appeared in print. Then on July 13, 1848, a young mother and housewife, Elizabeth Cady Stanton, had tea with four friends. She shared her discontent with the inferior status of women despite the revolutions that had promised equality in America, Britain, and France.

The group placed a small ad in the local newspaper, the *Seneca County Courier*. They announced that, "A convention to discuss the social, civil, and religious condition and rights of women" would gather at the Wesleyan Chapel in Seneca Falls on July 19.

At the gathering, Stanton presented the "Declaration of Sentiments," a document modeled after the Declaration of Independence, asserting the equality of men and women and demanding equal rights for women. The convention drew over 300 attendees, including the prominent abolitionist Frederick Douglass. It was covered in many newspapers and sparked a national conversation on women's suffrage and equality.

6.18. Elizabeth Cady Stanton.

The momentum from Seneca Falls continued to build over the following decades. In 1869, Stanton and Susan B. Anthony formed the National Woman Suffrage Association. Its primary objective was to pass an amendment to the US Constitution granting women the right to vote. They were vilified as "nasty women" by their opponents.

Internationally, the women's rights movement gained traction as well. In 1903, Emmeline Pankhurst founded the Women's Social and Political Union in the UK. Their tactics, including hunger strikes and civil disobedience, captured global attention and inspired similar movements in other countries.

By the early 20th century, women's suffrage was a significant issue worldwide. In 1893, New Zealand became the first country to grant women the right to vote. Australia followed in 1902 and several Scandinavian countries in the following decade.

World War I highlighted the contributions of women to the war effort and further justified their demand for equal rights. The US passed the 19th Amendment to the Constitution in 1920, granting women the right to vote. Many other countries, influenced by the persistence and resilience of the women's suffrage movement, began to adopt women's voting rights.

By the mid-20th century, women's rights were regarded as fundamental human rights across the globe. After millennia of a status quo in which women were the property of men, *in just 100 years the human race changed its collective mind* about gender equality.

Twilight of Colonialism

The right of one country to conquer, rule, and subjugate another country was unquestioned throughout most of history. The ancient Egyptians, Greeks, and Romans all expanded their territories through conquest, viewing the subjugation and exploitation of other peoples as a noble pursuit. Texts like the Bible and the works of Herodotus and Julius Caesar extol these conquests and the integration of vanquished peoples into expanding empires. The writers had an unquestioning belief that they were bringing civilization, religion, and culture to "barbarians."

The 20th century marked the beginning of the end for colonialism, starting with the British withdrawal from India in 1947. The independence movement, led by figures such as Mahatma Gandhi and Abdul

Gaffar Khan, highlighted the moral and economic unsustainability of colonial rule.

6.19. Captain John Smith trading with members of an Indigenous group.

Gandhi's nonviolent resistance strategy galvanized global support and set a precedent for other colonies. Following India's independence, a wave of decolonization swept across Asia, Africa, and the Caribbean. Nations like Indonesia, Ghana, and Jamaica gained independence, dismantling centuries-old colonial structures.

Today, colonialism is virtually nonexistent as a formal political system. The remnants of colonialism persist in various forms, but direct control over territories by foreign powers has largely ended. The international community, through organizations like the United Nations, champions self-determination and condemns any attempt at colonial dominance. Though colonialism had been unquestioned since the dawn of society, *global attitudes flipped in a few decades.*

There are many other social issues, like the abolition of slavery and universal voting rights, about which we as a human family quickly changed our collective minds. Here are some examples from very recent history:

- Assisted suicide
- Animal rights

- Bullying
- Child protective services
- Civil rights
- Clean air and water access
- Death penalty
- Disability accommodations
- Domestic violence shelters
- Drug legalization
- Equal pay
- Equal protection under the law
- Extreme poverty
- Freedom of association
- Freedom of movement
- Freedom of speech
- Gay marriage
- Gun safety
- Hate crimes
- Hate speech
- Homeless shelters
- Human rights
- Human trafficking
- Illiteracy
- Minority rights
- Physical abuse
- Racism
- Ridicule
- Sex trafficking
- Sexual abuse
- Torture
- Universal healthcare
- Universal primary education
- Verbal abuse
- Veterinary care for pets

That's a long and impressive list, especially when viewed as a whole. These reforms are all recent and collectively reflect the *rapid emergence of an increasingly compassionate society.* When you zoom out from the single thread of today's news headlines with their blood-red violence and hazy confusion and look at the whole quilt, the big picture that emerges is that compassion has been increasing for the past 500 years of human history. It has been *accelerating for the past 100 years,* a trend that is even more pronounced since the turn of the century.

The Evolution of Compassion

When collective consciousness shifts, profound changes in religion, law, society, art, government, education, and culture follow. These shifts typically start with a few people, but once a critical mass is reached and collective culture ignites, entire civilizations can quickly evolve.

The data bear this out.[49] Worldwide, extreme poverty has declined by about 90% in the past century. During that same period, life expectancy has doubled. Since 1970, illiteracy has halved. The wealth of the average global citizen has tripled since 1980.

6.20. Most ancient societies had little use for the elderly and disabled. Modern societies take care of them.

Measured by consumption, the US poverty rate has declined by 90% since 1960. The proportion of the world's population without access to clean air and water has roughly halved since the turn of the century. Dozens of jurisdictions now allow assisted suicide, up from just one—the US state of Oregon—in 2000. Google searches for racist and homophobic terms have declined 80% since the beginning of this century.

What all of these shifts illustrate is an *increase in compassion.* Society began to prioritize the wellbeing of others, such as children, sick people, slaves, women, minorities, prisoners, people with disabilities, and terminally ill patients. Those in positions of power, often *without any direct benefit to themselves,* empathized with the less privileged and willingly surrendered power.

I immigrated to the US in 1977, when Jimmy Carter was president. His actions in office had a profound effect on the whole direction of US government policy and some serve as shining examples of compassion.

JIMMY'S BIG DECISION

President Jimmy Carter's declaration that human rights would be a central pillar of US foreign policy marked a pivotal moment in American diplomatic history. In a speech delivered at Notre Dame University on July 14, 1977, Carter asserted that "our commitment to human rights must be absolute," signaling a departure from traditional realpolitik and a bold assertion of moral leadership on the global stage.

6.21. Jimmy Carter in 1977 in his first year as President of the United States.

Carter's stance was ridiculed in the media and by others in government. Critics derided his approach as naive and idealistic, arguing that it undermined strategic alliances and national security interests. Senator Jesse Helms, a staunch conservative and chairman of the Senate Foreign Relations Committee, mocked Carter's policy as "sanctimonious shilly-shallying."

Internationally, leaders of authoritarian regimes perceived Carter's human rights agenda as interference in their internal affairs. They condemned it as hypocritical, accusing the United States of applying double standards in its global interventions.

Despite the ridicule and opposition, Carter remained steadfast in his commitment to human rights. His administration implemented policies that linked US aid and support to improvements in the human rights records of various countries. Though Carter was not reelected, his bold statement of values lived on as a foundational principle of US foreign policy.

Jimmy Carter's historic announcement that human rights would be a cornerstone of US foreign policy faced denunciation from critics who viewed it as impractical and detrimental to national interests. However, his principled stance introduced a moral element to international relations. The historical record has shown him to be one of the most admired American presidents.

The value of human rights championed by Jimmy Carter, despite all the obstacles and ridicule he faced, is just one of many examples of how compassion is increasing in human societies.

Research shows that globally, human beings are experiencing a surge in spiritual, physical, and material wellbeing, despite what the news suggests. An analysis of word frequency in media across 130 countries since 1979 shows a rise in negative terms like "horrific" and "terrible" while positive words like "nice" and "good" have declined.[50]

The media's portrayal contradicts the reality of increasing human flourishing, which aligns with significant global changes in consciousness.

The recent past has seen humanity reassess issues like bullying, racism, homelessness, and civil rights. These changes often *stemmed from empathy rather than self-interest.* In Chapter 7, we'll see just how extensive these changes have been.

These aren't just shifts in global attitudes and behavior. They correlate with changes in the hardware of the brain. Scientific tools like MRIs and EEGs allow us to observe how *shifts in consciousness restructure the brain.* We now understand that changes like increased compassion lead to alterations in brain structure.

This makes it likely that *human brains have been evolving over the last 500 years.* The increase in compassionate behavior that we observe globally might have had the consequence of changing the physical structure of the average human brain. These significant changes in social structures might serve as external indicators of deep shifts in our collective consciousness. That force may be *changing the anatomy of our brains* on a global scale.

Compassion as a Driver of Brain Evolution

Was there a surge in meditation 500 years ago? Not that we can measure. However, there was a consistent rise in social reforms indicating increased compassion. In my book *Bliss Brain,* I review the MRI research on different types of meditation. This shows there is a single component of meditation that changes the brain more rapidly than anything else.

That factor is *compassion.* Activate the Empathy Circuit of Chapter 1 and you grow the neural networks of SQ *more quickly than any other type of meditative practice.*

Compassion might have been *gradually altering human brains for centuries.* Although we can't perform MRI scans on historical figures like abolitionists or suffragettes, the rapid changes we've seen in social norms imply large-scale brain evolution. As societies have grown even more compassionate in recent decades, brain evolution may have accelerated.

If this hypothesis holds, social change will continue to accelerate as compassionate brains foster even more compassion. Human society could transform dramatically in a few years, leading to the most compassionate and enlightened civilization in history.

The notion that emotions like compassion drive evolutionary change is not new. Charles Darwin explored this in his 1871 book *The Descent of Man*.[51] He wrote: "sympathy will have been increased through natural selection; for those communities, which included the greatest number of the most sympathetic members, would flourish best..."

We develop compassion early in life; children as young as 18 months respond to the distress of other people with appropriate facial expressions, gestures, or vocalizations.[52]

Evidence has been found that even early species of *homo* like the Neanderthals practiced compassion. In one burial site, the remains of a child with Downs syndrome was found alongside other members of the band. The skull revealed that the child was deaf and likely unable to walk or balance, but nonetheless lived till the age of six. This would not have been possible without compassionate care by other members of the Neanderthal tribe.[53]

6.22. Natural selection supports compassion and altruism.

A comprehensive scientific review identified multiple ways compassion influences evolution. It suggests that our distant ancestors "likely preferred mating with more compassionate individuals—a process that over time would increase compassionate tendencies within the gene pool."[54] Compassion also offers an evolutionary edge by *promoting cooperation with people who are not blood relatives.* Societies that foster compassion tend to thrive. This process might have been steering evolution since prehistoric times.

We aren't just *passively experiencing* a radical evolutionary shift. With our newly compassionate brains, we might be *actively creating* one. In Chapter 7, we examine the implications of this hypothesis across various fields, including education, medicine, law, science, business, technology, and art.

Compassion Changes the Brains of Individuals

Compassion changes the brains of meditators in many ways. In the MRI study I performed with colleagues Peta Stapleton and Oliver Baumann of Bond University, we found that the center of the Selfing Control Circuit of Chapter 1, the medial prefrontal cortex, became quiet. The left dorsolateral prefrontal cortex, the hub of "left-brain" rational activity, took a nice relaxing holiday.

But the insula, center of the Empathy Circuit and its associated emotions of compassion, gratitude, awe, and joy, lit up in a bright blaze of intensity. These functional and structural changes in the brain occurred after just 28 days of practicing EcoMeditation daily.[55]

In the brains of long-term meditators, the nucleus accumbens, which is active in cravings and addictions, begins to atrophy from disuse. As meditators light up the Enlightenment Network, the ventromedial prefrontal cortex, key to the Emotion Regulation Circuit, gets bigger and stronger. Ditto the orbitofrontal cortex, central to the Attention Circuit.

The activation sensitivity of the amygdala, the brain's "fire alarm," declines in experienced meditators. A study comparing monks who have meditated for an average of 19,000 hours to a group with 44,000 hours found "a staggering 400% difference in the size of the amygdala response" between the two groups. This indicates that *the brain's stress networks keep contracting throughout the meditator's life.*[56]

The thickness of gray matter helps scientists measure the biological age of a brain, as distinct from its chronological age, as we saw in Chapter 1. Brains with active Enlightenment Networks stay more youthful, with more cortical tissue than Caveman Brain.

This research answers the crucial question of how much change you can create in your own brain. If you exercise your superpower to create new neurons consistently, there's no upper limit. *Compassion can keep evolving your brain throughout your entire life.* And the inner light you experience is just the beginning. Patanjali's *Yoga Sutras* talk about "radiant beings" whose light extends beyond their bodies, as exemplified in the life of Martin of Porres.

SAINT MARTIN AND THE RADIANT LIGHT

Born in 1579 in Lima, Peru, Martin of Porres was the son of a Spanish nobleman and a freed slave of African descent. Raised in poverty, he grew up facing discrimination and injustice at every turn.

6.23. St. Martin de Porres.

At the age of 15, he volunteered at the Church of Holy Rosary, run by the Dominican order. Despised because of his mixed-race "mulatto" heritage, he was set to work in the lowest station available, scrubbing pots and pans in the monastery kitchen.

Martin's compassion knew no bounds. He fed stray animals. He shared his meager rations with the poor, regardless of his own hunger. He even fed the rats and mice in the monastery kitchen, seeing in them creatures deserving of care.

Legend has it that Martin performed a miraculous healing of a dying man through his deep faith and prayer. As the man lay near death, Martin knelt beside him, fervently praying for his recovery. Witnesses recounted that during his prayers, the room filled with a radiant light, and the man's health swiftly improved.

One day, Martin encountered a beggar lying in the street, crippled and unable to move. Without hesitation, he lifted the beggar onto his own shoulders and carried him to shelter.

Eventually, the monks relented and allowed him to join the order. Martin founded an orphanage and a hospital where he cared for the sick, regardless of their circumstances. His humility and devotion led him to treat all people equally, reflecting his belief in the inherent dignity of every person.

During the Vatican's investigation into his life and virtues, it was noted that Martin exemplified Christlike compassion. The inquisitors wrote: "He showed God's love through his acts of mercy towards all those in need," and "His life was a living example of charity and humility, inspiring countless souls to follow his path."

Today St. Martin of Porres is recognized as the patron saint of social justice, mixed-race people, public health workers, and those seeking racial harmony.

St. Martin's story reminds us that we don't have to possess great resources to exemplify SQ. Martin managed that feat with only the modest congregation of the rats and mice he encountered in the monastery kitchen. When we live from a core of compassion, we make a difference in the small circle of life around us. But we're part of a larger group of compassionate people, and when we consider the cumulative effect, we recognize that compassion *is a powerful force for massive social transformation.*

With Rising Numbers Becoming Compassionate

What percentage of the world's population is actively developing their SQ? How has this proportion been changing over time?

Records are available from sources such as the National Center for Health Statistics in the US and the European Values Study in the EU. They indicate that during the 20th century, the number was around 1%.

This 1% appears to have been *a baseline number for hundreds of years in many different societies.* After Duke William of Normandy conquered England in 1066 CE, he sent officials all over the country to compile a survey of his new domain. They found about 33,000 people engaged in various spiritual pursuits. That represented about 1% of the population at that time. Similar percentages are found in records from medieval France and Germany, as well as ancient India and China.[57]

By 2012, US government records indicate that the number of meditators had grown from 1% to 4%.[58] That's a fourfold increase in around 20 years. By 2017, it had exploded to 14% and has continued to rise from there, likely *exceeding 20% of the population* today. That's a big dose of freshly minted compassion circulating through the world, bringing with it the SQ to change both other people and society.

Almost Half the Population Experiences Altered States

SQ is also evident in the high prevalence of altered states of consciousness. A large-scale study that included 3,135 adults in the US and the UK found that *45% had experienced altered states of consciousness* catalyzed by meditation, mindfulness, or yoga.[59] This was a much greater percentage than the investigators were expecting to find.[60]

"Altered states were most often followed by positive, and sometimes even transformational effects on wellbeing," observed study author Matthew Sacchet, professor of psychiatry at Harvard Medical School.[61]

Among the experiences reported were "derealization"—a feeling of being detached from the world around you—as well as orgasm-like ecstatic thrills, heightened sensory perceptions, feelings of heat or electricity running through the body, perception of inner light, out-of-body experiences, and nonduality, the sense of unitive oneness with All That Is. Those are among the key markers of transcendent states identified by researchers in Chapter 4.

Such experiences are much more common than we might suppose. Another large-scale study of meditators found that *80% had experienced the most dramatic of altered states,* such as "loss of awareness of where you are," "experience unity with ultimate reality," and "ecstasy."[62]

When people who've experienced Bliss Brain come down from the mountaintop, what do they do next? Many of them are motivated to share their wisdom with others who are suffering and develop an interest in *seva* or sacred service. Compassion drives them to devote segments of their time and gifts toward helping other people.

Rise in Nonprofit Formation

One proxy for compassion on a national scale is the number of nonprofit charities in existence. These are formed in response to a variety of human needs, from feeding the hungry to housing the shelterless to educating children to caring for the aged. Forming and running a nonprofit is a great deal of work, as I know from many years experience with the National Institute for Integrative Healthcare and the Veterans Stress Solution, both of which I founded in 2007.

Forming a new nonprofit is an uphill slog. It involves obtaining funding, persuading busy people to serve on your board of directors, meeting government standards, creating procedures, holding meetings, designing programs, and daily management. This presents considerable barriers to entry for new charities.

Yet thousands of groups of dedicated people climb this mountain every year. The number has been growing steadily, as compassion drives people to prioritize the welfare of others. In 1990 and earlier years, about 20,000 new nonprofit charities were formed in the US annually. But after that, the number began to rise. By 2000, some 50,000 were being formed in the US each year.[63]

Today that number has risen to *100,000 annually.*[64] That's a fivefold increase in the intention of people and groups to do good and help others. It demonstrates that *we're a much more compassionate society than we were in the past century* and that the impulse to altruism is growing.

Compassion Changing Brains on a Large Scale

MRI research has allowed us to peer into individual human brains and determine that meditation changes the brain. It's shown us the brain regions involved and how they work together as the Enlightenment Network. It's demonstrated that key circuits grow, some slowly over time, others in as little as 30 days. It shows that sustained meditation continues changing the brain even into old age. We also know that the number of meditators has grown 20-fold in the last few decades.

This means that *the brains of millions of people today are anatomically different from that of their parents.* This structural brain change is related to compassion. We see the rise in compassion in the collective culture of humankind over the last 500 years and consider the possibility that this may be an indicator of brain change in millions of people all over the world. When I was writing the final chapter of *Bliss Brain* in 2018, the evidence from social change led me to hypothesize on page 278 that this might be producing anatomical change. Now, experimental evidence is emerging to support this hypothesis.

What might such brain change look like and how can we identify it through research? Intriguing studies are now giving us clues about how *brains may have been changing on a large scale in the past century,* how those changes are linked to *the regions active in meditation,* and how the process might accelerate in the future.

HUMAN BRAINS ARE GETTING BIGGER

Research reveals an intriguing trend: *human brain sizes have been on the rise over the past few decades.* People born in the 1970s boast a 6.6% larger brain volume and nearly 15% greater cortical surface area compared to those born in the 1930s.[65]

This enlargement of the brain could play a crucial role in enhancing our brain's resilience, reducing our susceptibility to age-related dementia, and boosting SQ.

With a higher life expectancy, a larger portion of the population of developed countries faces the risk of Alzheimer's disease and similar dementias. Data, including that from the long-running Framingham Heart Study, have shown a decreasing trend in the incidence of dementia.

This sparked curiosity among researchers about whether early life enhancements in health, education, and neurological risk factors might lead to improved brain development and increased brain size.

The Framingham Heart Study began in 1948 with a comprehensive examination of 5,209 adults—virtually the entire population of Framingham, Massachusetts. Originally aimed at understanding cardiovascular health patterns, the study broadened to include brain health, happiness, and many other metrics. It benefits from data collected from several generations of residents.

For this particular study, researchers analyzed high-resolution MRI scans from participants born from the 1930s to the 1970s. These data covered a diverse demographic, capturing many age groups with a wide variety of lifestyles and health practices.

The study examined 3,226 participants, nearly evenly split between females and males, all of whom underwent MRI scans to provide detailed insights into their brain structures.

The analysis of individuals born in different decades revealed a clear trend: an increase in brain volume and cortical surface area in later generations.

Average skull capacity increased from 1,234 mL for those born in the 1930s to 1,321 mL for individuals born in the 1970s. Similarly, the average cortical surface area expanded from 2,056 square centimeters for the earlier generation to 2,104 square centimeters for those born in the 1970s.

Among the changes were *increased volume of the hippocampus.* This structure includes the dentate gyrus, which we saw in Chapter 1 plays a key role in the Emotion Regulation Circuit.

Brain size increases were significant even after adjustments for body height and other factors, suggesting that factors beyond genetics, like health, social, cultural, and educational influences—all part of collective culture—contribute to brain size and health.

"Genetics plays a major role in determining brain size, but our findings indicate external influences—such as health, social, cultural and educational factors—may also play a role," noted the study's lead author, neurologist Charles DeCarli. "A larger brain structure represents a larger brain reserve and may buffer the late-life effects of age-related brain diseases like Alzheimer's and related dementias."[66]

The *larger hippocampus of the more recent brains* is especially significant. A bigger hippocampus has been identified as a characteristic of meditators.[67] It assists them with enhanced memory, learning, and emotion regulation. It's one of the key components of the Enlightenment Network. Having sufficient neural mass to dial down negative emotions like anger, fear, resentment, guilt, shame, blame, and overwhelm is crucial to cultivating SQ. As a bonus, your chances of dementia diminish.

Evolved Brains Connect

There's also intriguing research showing that evolved brains resonate *with each other and with larger structures in nature and the universe.*

One of the experimental designs that illustrates *resonant nonlocal communication* is for two people to meditate in the same room together for 20 minutes. They are then sent to separate electrically shielded rooms where their brain activity is measured. The rooms are constructed to eliminate the possibility of any known energy entering or leaving. This rules out electromagnetism, acoustics, light, and other possible mechanisms of communication between participants.

Despite this separation, their brain-wave patterns begin to sync. When a light is shone in the eyes of one of them, the brain waves of the other respond. After a series of these trials, one set of researchers concluded, "The transferred potentials demonstrate brain-to-brain nonlocal EPR correlation between brains, supporting the brain's quantum nature at the macro level."[68]

Andrew Newberg found that the effect scales, with the size increasing the greater the number of people with awakened Enlightenment Networks. People who pray together enhance the rate at which incoming worshippers activate the neural pathways of SQ. Their brains are entrained together.[69]

With the evidence of increased compassion emerging worldwide over the past 500 years, it's possible that *large numbers of brains might be communicating in resonant global entrainment.* Human beings may be coevolving, sparking the evolution of SQ in each other.

Resonance and Patterns in Nature

The Earth itself has a primary frequency of 7.8 Hz or cycles per second. This is also one of our six primary brain waves. It is in the frequency band called theta, between 4 and 8 cycles per second, which we examined in Chapter 5. During sleep, our brains are in the slower band of delta for most of the night. But for brief periods every hour or so, they speed up to theta. Our eyes move rapidly in their sockets even though our eyelids are closed, and we have vivid dreams.

During these rapid eye movement or REM periods of sleep, neurons in the hippocampus fire furiously. As we process information in our dreams, they wire together. Theta REM sleep is when most new synapses are built in the brain's memory, learning, and emotion centers.

Healers are also in theta during the healing process. Though their brain waves may be all over the place when they're in ordinary states of consciousness, when they enter the healing space, they drop into theta, as we saw in Chapter 5. Specifically, their brains fire at 7.8 Hz, *resonating with the frequency of the planet itself.* This resonance may amplify the healing effect.

We're resonant with the physical world in many other ways as well. The delta waves produced by our brains have the same frequency as the waves of the ocean. Delta also resonates with the waves produced by the fluctuation of the Earth's geomagnetic field. Chanting "om" produces a frequency around 110 Hz, which induces coherence in our heartbeat. Our heart's electromagnetic field, the strongest of any organ in our body, then entrains our brain into coherence as well.

This type of *coherent* resonance is found *all over nature* too. Birdsongs in the 2–4 kHz range can influence seed germination by resonating with structures in the cells of plants that respond to acoustic signals. Cicadas produce mating calls at frequencies around 4–6 kHz, resonant with natural frequencies found in tree branches and leaves. This amplifies the calls of the insects, enabling them to attract mates over long distances.

The wingbeat frequency of bees, typically around 250 Hz, can resonate with the frequencies of flower petals to help release pollen. Wolves

6.24. Fractals are patterns that repeat at large and small scales.

use howls with frequencies between 150 Hz and 900 Hz. These resonate with the natural acoustics of forests, allowing their calls to travel long distances through the dense vegetation, facilitating communication among members of the pack. Elephants communicate using infrasound frequencies below 20 Hz. Because these resonate with the Earth's surface, they travel through the ground, allowing communication with other herd members several kilometers away.

Similarly, the universe is full of repeating patterns. The ratio of the universe's electric force to the gravitational force is approximately 10^{40}. The ratio of the size of the universe to the size of elementary particles is also 10^{40}. The number of nucleons in the universe is roughly equal to the square of 10^{40}. After reviewing many numbers in physics at both the macro and micro levels, including the gravitational constant, the charge of electrons, the speed of light, the mass of elementary particles, and microwave radiation, esteemed scientist Ervin Laszlo says that "the physical parameters of the universe turn out to be extraordinarily correlated and coherent."[70]

Universal Consciousness and the Human Brain Interact

The journal *Popular Mechanics* has posited that human consciousness can interact with the universe. It cites experiments suggesting that consciousness might exist as a quantum wave, capable of connecting with the cosmos. The article further explains that conditions in the brain are "just right" to support consciousness as a wave that can interact with the universe.[71]

One mechanism by which *Popular Mechanics* thinks this might occur is resonance between the electromagnetic fields of our brains and the universe in which we live, especially the sun. A series of experiments has examined the correlation between human heart rate variability and the variations in Earth's magnetic field triggered by solar flares. Over the course of 30 days, the two track each other, as can be seen in the following graph.

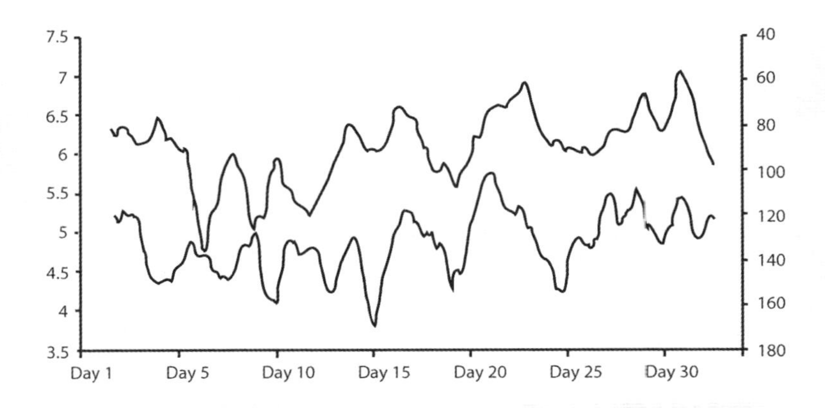

6.25. Top: Earth's magnetic field variations (total power) over the course of a month. Bottom: Human heart rate variability (high frequency power) over the same period.[72-73]

Not only are the cosmic and human systems energetically linked, they are structurally similar too. Research has revealed close parallels between them. When examined closely, *the structure of our brains looks remarkably like that of the universe*, though at a much smaller scale.

THE STRUCTURE OF THE HUMAN BRAIN LOOKS LIKE THE UNIVERSE

The structure of the universe and neural networks in the human brain share striking similarities. They were mapped by astrophysicist Franco Vazza and neuroscientist Alberto Feletti working in close collaboration. They conducted an extensive series of numerical assessments of the two systems and found them to be astoundingly alike.[74]

Using a combination of techniques from cosmology, neuroscience, and network analysis, Vazza and Feletti made a quantitative comparison between them. Despite the vast difference in scale, these two complex systems show remarkable similarities.

They found that the brain resembles a "3-pound universe" with remarkable correlations between neural networks in the brain and the cosmic web of galaxies.

The brain contains about 100 billion neurons that form around 100 trillion connections. These neurons are organized into a hierarchical structure of nodes, filaments, and clusters—all shaped by our thoughts, feelings, and emotions.

The observable universe has roughly 100 billion galaxies. The interplay between gravitational forces and the universe's accelerated expansion creates a cosmic web of filaments composed of ordinary and dark matter. Galaxies cluster at the intersections of these filaments, with empty spaces between them, closely mirroring a neural network.

Scientists estimate that only about 25% of the universe's matter is visible, with the remaining 75% being dark matter. Similarly, neurons make up less than 25% of the brain's mass, with the rest being non-neural tissue and water.

"Although the relevant physical interactions in the above two systems are completely different, their observation through microscopic and telescopic techniques have captured a tantalizing similar morphology, to the point that it has often been noted that the cosmic web and the web of neurons look alike," noted Vazza and Feletti.[75]

6.26. The human brain and universe display striking structural similarities.

To delve deeper into these similarities, the researchers used power spectrum analysis, a technique often used in astrophysics to study the large-scale distribution of galaxies. This method allowed Vazza and Feletti to measure the strength of tiny fluctuations across various spatial scales in both a simulation of galaxies and sections of the brain's cerebellum and cerebral cortex.

"Our analysis showed that the distribution of the fluctuation within the cerebellum neuronal network on a scale from 1 micrometer to 0.1 millimeters follows the same progression of the distribution of matter in the cosmic web but, of course, on a larger scale that goes from 5 million to 500 million light-years," observed Vazza.[76]

The team also compared the power spectra of other complex systems, such as tree branches, clouds, and water turbulence, but none exhibited the same degree of similarity as the brain and the universe.

However, power spectra alone do not provide insights into the complexity of these systems. To address this, the scientists examined the networks of both systems, focusing on the average number of connections per node and the clustering patterns of these nodes.

"Once again, structural parameters have identified unexpected agreement levels. Probably, the connectivity within the two networks evolves following similar physical principles, despite the striking and obvious difference between the physical powers regulating galaxies and neurons," said Feletti.[77]

Since the human brain evolved on planet Earth spinning around our sun in this universe, it is not hard to understand that *similar patterns can be found in all of them.* Human consciousness evolved in the field of this universe, so it makes sense that they resonate together. Laszlo observes that "human awareness and intention have their roots in the cosmos; they were there in potential at the birth of the universe."[78]

Physics agrees on four fundamental forces. These are electromagnetism, gravity, the strong nuclear force, and the weak nuclear force. The strong nuclear force binds protons and neutrons together in the nucleus of an atom. The weak nuclear force is what results in radioactive decay. In my book *Mind to Matter,* I review experiments showing that *human consciousness can affect all four of these forces,* indicating that consciousness underlies them all.[79]

Coevolution of Universe and Human Brain

In Chapter 1, I defined SQ as "the ability of human consciousness to interact with universal consciousness."

Mystics in that state of connection talk about "being one with everything" and their brain function indicates that this is a literal description of their experience. The first of the five characteristics of transcendent experiences from Chapter 4 was "a sense of oneness with the universe and nature." The third was "a perception that mystical experiences are more real than everyday reality."

What science is showing us is that these brain-to-universe connections aren't just evocative figures of speech. *The human brain and the universe may be coevolving* in a two-way dance, with millions of meditators "tapping in" to the universe, which is, in turn, evolving the structure of their brains.

Laszlo reminds us, "Matter and mind are not separate, distinct realities; they are complementary aspects of the reality of the cosmos."[80] He goes on to state that "In a higher state of consciousness...we enter into deep communication with the universe. In these states the awareness of every cell in the body coherently resonates" with the universal field.

6.27. In transcendent states, we resonate with the universal field.

We Live in a Conscious Universe

Drawing on hundreds of studies, Laszlo finds that *consciousness is the fundamental property of the universe itself.* In transcendent states, SQ allows us to access this consciousness. What we experience there is an "immense and unfathomable field of consciousness endowed with infinite intelligence and creative power."[81]

The universe is itself conscious. At the most irreducible level, beneath the four forces of physics, it is composed of consciousness. In the mystical experience, human consciousness participates in universal consciousness.

It is SQ that harmonizes individual consciousness with the consciousness of the universe. It *brings human awareness into resonant connection with the fields of the cosmos.* This resonance in turn alters the anatomy and physiology of the brain that enters the state of communion with the universe.

When we elect ourselves into the ranks of evolutionaries by meditating and entering transcendent states, we're entering a resonant cosmic field. All other people in similar states are also in this field and the number of people doing this has risen 20-fold in the past half-century. Collective SQ is accelerating social change toward a compassionate and connected civilization. In Chapter 7, we'll see what this looks like.

Full-Speed Evolution

Both individual and social evolution are unfolding at a pace that would make Charles Darwin's head spin. The Cambrian Explosion—so named because evolution proceeded much more quickly than usual during this period—took 12 million years. Then the human brain tripled in size over the three million years between the creation of the first knapped tool and the smartphone—a rate that is even swifter by evolutionary standards.

The theory of punctuated equilibrium takes the interval required for rapid evolution down to 50,000 years. Major social shifts used to take thousands of years, then hundreds; now they can happen in less than a decade.

Which leads us to you and me. Here you sit, a master manifester, capable of *inducing significant anatomical changes in your brain's structure in a mere eight weeks.*

This phenomenon is unprecedented in evolutionary history. No species before *Homo sapiens* has possessed the ability to *alter the hardware of its brain using the software of its mind*—and do this not over millennia but in weeks. This unique human ability is *changing the entire course of evolution.* In the next chapter, we'll discover the kind of world this will usher in.

At the same time, *brain-universe co-creation is resulting in a progressively more compassionate world.* At times that's hard to see because of all the current fear-based news stories clouding the picture; you have to zoom way out to the 30,000-foot view in order to spot the real patterns. But it's possible to extinguish even the highly conditioned fear responses that have kept human beings trapped in the same negative behaviors for millennia. Millions of meditators are changing their brains in precisely the ways required to think and act unlike their ancestors.

The combination of brain change and universal connection is *guiding evolution on a completely new path.* Extrapolating from the patterns of the past will not show us the trajectory of the future. We are in uncharted waters, as individual SQ resonates with the SQ inherent in the universe to co-create a compassionate new reality.

Deepening Practices

The Deepening Practices for this chapter include:

- **Journaling Exercise 1:** In thinking back on your life so far, are there any times that you took an "evolutionary leap," that is, made a radical change in your circumstances, consciousness, or behavior? Describe each and write about the results of that change in your life.
- **Journaling Exercise 2:** Think about the role compassion has played in your life. Describe some pivotal events when you received compassion from another and when you were the one extending compassion. Consider and write about how each made you feel.
- **Dawson Guided Neuroliminal:** Activating Self-Compassion

Get the meditation track at: SpiritualIntelligence.info/6.

Extended Play Resources

The Extended Play version of this chapter includes:

- BBC video of catfish catching pigeon
- Podcast: Exploring Consciousness: Ervin László and Dawson Church in conversation
- Podcast: The Awakened Brain: Lisa Miller and Dawson Church in conversation

Get the extended play resources at: SpiritualIntelligence.info/6.

Updates to This Chapter

As new studies are published, this chapter is regularly updated. Get the most recent version at: SpiritualIntelligence.info/6.

Chapter 7

Accelerating Human Flourishing

Brain evolution is combining with cumulative culture to produce a progressively more compassionate society. The evidence from social change as well as biological evolution points towards an unprecedented era of human flourishing in the coming years.

You and I can witness Spiritual Intelligence both driving the process and being influenced by it. Compassionate human brains and minds connect with universal consciousness and in turn are shaped by it. Changes of mind now spread from a few to the many at an unprecedented speed. Eventually the tipping point is reached for an entire society.

With hundreds of millions of people changing their own brain anatomy, what large-scale effects can we anticipate in the time ahead? What kind of world does the enlightened collective produce?

The ripples will spread to every field of human endeavor, changing many of them beyond recognition in just a few decades. In the following pages, we'll examine specific fields in which SQ is likely to change our individual behavior and collective future.

The Trend of Global Wellbeing

Though we can anticipate a better world as SQ spreads, we're building on a base of global wellbeing that shows a centuries-long pattern of

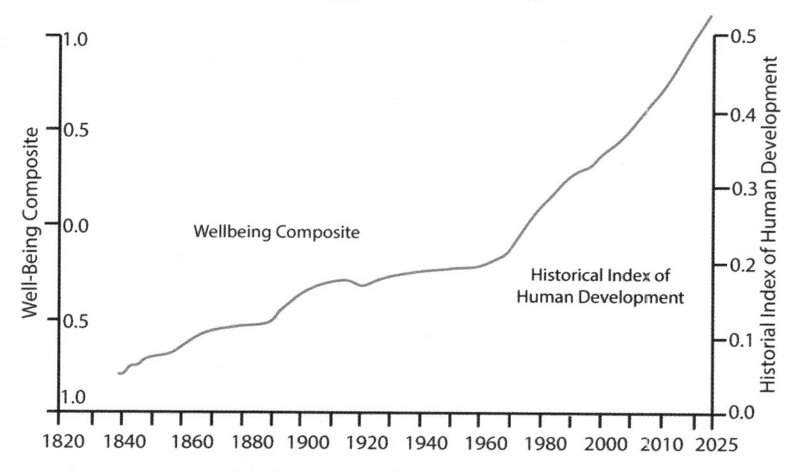

7.1. Global wellbeing from 1820 on.[1,2,3]

This trend line uses the Wellbeing Composite from 1820 to 1940 and the Historical Index from 1940 onward.

improvement. Global wellbeing has been increasing for the past two centuries, when measured worldwide with big data.

The trend line in this chart is a composite of many key measures. These include health, longevity, income, education, democracy, homicide, and biodiversity.[4] Don't let the simplicity of this graph fool you. It's a composite of billions of data points, hundreds of scientific studies, and meticulous research methodology. Yet it shows in just one simple chart that, decade by decade, *human wellbeing has been advancing*.

You might wonder about the poorest regions of the world. Does growth in the large industrial economies mask declines in the rest?

Though the regions of the world may be highly unequal, all of them have been improving at one rate or another. The worst-off parts of the world today are superior to the best-off parts of the world less than a century ago. If the world were divided between the West and "the Rest," the Rest by 2007 had reached the same level of wellbeing as the West enjoyed in 1950.[5]

There are elements of SQ in all these components of wellbeing. Compassionate people take care of others and the environment. They enjoy better health and greater longevity. Though it's not possible to chart SQ or brain evolution using direct measures like those on the chart above, the correlation between SQ and these factors is striking.

This chapter will show us what science predicts about all the dimensions of human activity that went into this chart, as well as other key areas of human endeavor. As SQ continues to increase, it's likely to shift the human and planetary landscape in many different ways and further extend global wellbeing.

Counter-Trends

First, though, let's look at the counter-trends. While humanity is experiencing a "great flourishing," many people do not see things that way. Bleak and dystopian visions of the world are rife in politics, news, and entertainment media.

This has contributed to the unforeseen emergence of populist movements and leaders in the 21st century. They often champion causes—racism, fossil fuels, divisiveness, polluting industries, misogyny, political repression, intolerance, nationalism, homophobia, slash-and-burn agriculture, social conflict, adversarial policies, international rivalry, territorial conquest, bigotry, cynicism, anticompetitiveness, xenophobia, and so on—that the late 20th century seemed to have buried.

Besides the counterfactual bogeymen conjured up by unsavory political movements, we as a planetary species face real and objective challenges, some of which threaten our collective survival. These include wars, catastrophic climate events, economic shocks, health threats, technological disruptions, environmental pollution, pandemics, weaponized artificial intelligence, mass extinction, and the ever-increasing pace of change.

When I share the big-picture trends about human flourishing with friends, they often respond with incredulity. They point to the news stories about all these stark realities and can't believe the world isn't going to hell in a handbasket.

Yet crises have always been present. If you dig into the archives of any news outlet from any day of any year in the past, you'll find similar upheavals. What distinguishes today is the heightened degree of worry people have about them.

My informal observations are backed by data. Two British psychologists analyzed the words used in over 150,000 popular songs released over a 50-year period. They found a steady decline in words with a positive emotional valence, such as "love," "happy," and "joy." Concurrently, words with a negative emotional valence, such as "pain," "misery," and "sorrow," gradually increased. The word "hate" was virtually absent in the early years but surged dramatically after the turn of the century.[6]

News Headlines Grow Steadily Darker

An analysis of news headlines reveals a similar trend. Researchers examined data from 23 million stories published in the first two decades of this century, tracking occurrences of Paul Ekman's six basic emotions (anger, disgust, fear, joy, sadness, surprise) and a seventh category of "neutral" words. They found a steady increase in headlines with negative sentiments like anger, disgust, fear, and sadness, and a decrease in "neutral" sentiments. This trend was evident in both left-leaning and right-leaning media outlets.[7]

The General Social Survey conducted by the National Opinion Research Center (NORC.org) monitors ongoing trends in opinions and attitudes in the US. It found that between 2000 and today, the percentage of people categorizing themselves in the lowest happiness category more than doubled.

Despair in the Bottom 20%

This trend of declining happiness is not confined to the world's richest countries. A Gallup survey of people in 140 countries reveals a similar trend. It asks people to rate their lives from 0 to 10, with 10 representing the best possible life and 0 the worst. At the turn of the century, only 1.6% of people worldwide rated themselves at 0. Today, that number has quadrupled.[8]

Misery has increased notably in large countries like Mexico, Brazil, India, and China. In India, the percentage of people rating themselves at 0 has risen to 21%.[9] People were particularly unsatisfied with their work

7.2. The majority of people are unsatisfied at work.

situations. Gallup found that only 20% of people are thriving at work, with 18% hating their jobs and 62% feeling indifferent.

Flourishing in the Top 20%

Yet the Gallup survey also finds that while the bottom 20% of global citizens are experiencing the lowest-ever levels of wellbeing and happiness, the top 20% are reporting the highest-ever levels.

They're describing unprecedented levels of "flow" and self-transcendence. SQ lifts their consciousness out of preoccupation with daily life as they merge with that greater universal consciousness. Going to Maslow's state of "self-transcendence," many catapult their awareness into states of bliss and wellbeing unimaginable to the bottom 20%.

SQ is independent of socioeconomic status. Many of the fMRI studies covered in earlier chapters have Tibetan monks as their subjects and find them to be extraordinarily happy. Yet the monks own nothing, not even the saffron robes on their backs, which revert to the monastery upon their death.

Studies of Finders who've awakened to self-transcendent states find them at all levels of physical and financial wellbeing. Once they awaken, they usually remain awake despite the ups and downs of health, work, relationships, and money.[10]

The Awakeners

My research puts me in personal contact with many of these people. I lead regular "Awakeners" groups with people who are going through our "Short Path to Oneness" program. The term "Awakeners" aptly describes people who have risen in consciousness above the self-actualized top of Maslow's hierarchy.

The Awakeners live in the same world as the unhappy people. They read the same news and deal with the same economic, climate, family, and work realities. Yet they tend to be extraordinarily happy amidst it all. SQ leads to a sense of connection with that universal consciousness greater than the local ordinary self and it persists amongst the chaos of the world.

Awakeners also *set up their lives to support this sense of wellbeing*. They usually meditate, spend time in nature, connect on a soul level with close friends, choose meaningful work, direct their attention toward the good, consume positive media, and maintain that rock-solid sense of what Jeffery Martin calls "fundamental wellbeing." This sense of wellbeing underlies every aspect of one's personal existence, forming a core assumption about the world, others, and oneself.

The Great Divergence

Our world civilization seems to be experiencing a bifurcation of consciousness. Those who suffer are unhappier than ever before, often perceiving the world as worse than it has ever been, even though objectively, global wellbeing is increasing.

The Awakeners, however, live with an unshakable sense of wellbeing. As we saw in Chapter 6, about 1% of people in the world were meditating in 1980. That number exploded to 14% by 2017 and today approaches 20%.[11] Some 40% of Americans engage SQ through practices like meditation and prayer at least once a week.[12] The numbers are similar for other countries that keep official records.

7.3. Awakeners dancing at an EcoMeditation retreat.

These trends show that the number of Awakeners is increasing. While many people are suffering more than ever before and perceiving the world through the lens of negative emotion, the number of Awakeners is also growing. Objectively, the two groups live in the same world, but subjectively, their experiences have diverged dramatically.

This paradox illustrates the power of consciousness. Every human being has 86,400 seconds in each day, and in each moment, we can direct our awareness in any direction we choose. As a species and as individuals, we need to take action on real problems like climate change, social inequality, wealth disparity, food adulteration, racial bias, human rights violations, interpersonal violence, government corruption, medical inequity, gender inequality, and corporate greed.

However, it's equally true that we live in a world that offers experiences of serenity, beauty, awe, and grace every day. Both realities coexist. Neither negates the other. We can awaken and live self-transcendent lives amidst the perpetual upheaval and chaos that have always characterized the world. We can also lift our heads above the headlines of the day and read the data. The data show us that *the optimistic view is the one most aligned with objective reality.*

Where Spiritual Intelligence Is Taking Us in the Future

Despite the problems that beset our species, science shows us that global wellbeing and SQ are both rising. Hundreds of millions of people are meditating, rewiring their brain anatomy through compassion. This is producing large-scale global effects, with more to come.

Cumulative culture and rapid evolution will transform every field of human endeavor—some beyond recognition—in the next few decades. Here, we'll examine specific areas where SQ is poised to influence both our individual behaviors and our collective future.

Health

If you've ever been sick—and who hasn't—you'll agree that feeling physically healthy is central to happiness. How do we optimize our health? SQ is a key leverage point.

People with high levels of SQ enjoy healthier lives. The difference is apparent in many different measures of health.

Heart disease is the number one cause of death in the US, accounting for a quarter of all deaths. Research shows that mindfulness meditation improves all markers of heart health. It reduces blood pressure, improves performance on cardiovascular stress tests, lowers resting heart rates, and improves heart rate variability. The American Heart Association now officially recommends mindfulness as an adjunct treatment for the prevention and treatment of coronary disease.[13]

I examined physiological markers of health in one of my EcoMeditation studies. We found that just a single weekend of practice improved multiple markers of health.[14] In a related study, participants used EFT acupressure tapping to reduce stress, especially emotions arising from childhood trauma, as well as EcoMeditation. In just a week, their wellness indicators improved dramatically.

Their pain went down by 57%. Cravings for unhealthy foods reduced by 74%. Their resting heart rate, a measure of overall health, dropped 8%, and blood pressure by a similar margin. Their stress hormone cortisol reduced by 37%. That's a huge drop for so short a time. Simultaneously, their immune markers skyrocketed, with a 113% increase in a key molecule that neutralizes viruses entering the body.[15]

Meditation improves Alzheimer's symptoms, and in those already diagnosed with the disease, improves scores on cognitive tests, as we saw in Chapter 2. In healthy elderly people, meditation improves attention and mitigates cognitive decline. SQ is associated with an enormous array of health benefits.

JACK SCHWARZ AND THE HEALING POWER OF LOVE

Jack Schwarz was born in the Netherlands in 1924. When the Nazis occupied Holland, he was interned in a concentration camp, where he endured extreme physical and emotional torture. Yet he soon realized he could regulate the pain he was enduring. He began to pray and meditate, sharpening his mind to the point where he could continue his communion with the Infinite in the midst of horror.

Even in the bleakest moments, his SQ gave him a sense of deep connection to a higher power, which he believed was the source of his strength and healing abilities. "I knew that God was within me, and that His love could work miracles," Schwarz once said.

After the war, Schwarz dedicated his life to exploring and teaching the principles of self-healing. He demonstrated astonishing feats that defied conventional medical understanding, such as inserting large skewers into his arms or cheeks. The holes would rapidly close up without bleeding or pain. He could press lighted cigarettes into his skin without producing any sign of burning or injury. Under laboratory conditions at the Menninger clinic, he was able to replicate these feats of healing.

Schwarz attributed these abilities to his deep spiritual practices and unwavering faith. "Love is the greatest healer of all," he declared, emphasizing that it was through love—both for oneself and for others—that true healing occurred. He did not regard his abilities as exceptional, but believed that anyone could develop them.

7.4. Jack Schwarz.

Schwarz saw the human body not just as a physical entity, but as an integrated system that responds to our consciousness. Schwarz believed that by maintaining a strong spiritual connection, any of us can tap into our innate healing potential. His dramatic demonstrations showed millions of people worldwide that the body has extraordinary regenerative powers.

Many important genes are regulated by meditation, including those that stimulate the mitochondria in your cells to produce energy, that suppress inflammation, that control insulin production, and that dampen oxidative stress. Just eight hours of meditation are all that's required to begin boosting your immune system.[16]

As more and more people meditate, expect to see this reflected in declining statistics for many diseases. These include cardiovascular diseases, autoimmune diseases, and age-related decline.

SQ produces drops in cortisol and inflammation along with silencing of the associated genes.[17] Self-transcendent people improve the quality of their health, happiness, and overall quality of daily life as well as reducing their stress levels. The six primary genes involved are so important they have been preserved by evolution from primordial times and are found in life forms from single-celled organisms on up the evolutionary ladder to humans.[18]

When Alzheimer's patients supplement meditation with healthy lifestyle choices, the combination turbocharges their healing. Research shows that following a healthy plant-based diet, getting moderate exercise (like walking, stretching, and strength training), breathing exercises, imagery, and spending quality time with friends and family all contribute to reversing Alzheimer's symptoms.[19]

This is assisted by SQ, which correlates with greater physical health, prosocial behavior, resilience, and reduced burnout and stress.[20] For more on reversing Alzheimer's and cognitive decline naturally, see the Extended Play resources at the end of this chapter.

Lifespan

Most of us don't want just a healthy life; we want a long one. Meditation is key to longevity. It slows the aging of your cells, increasing your lifespan.[21]

Chains of molecules called telomeres form the end caps of chromosomes. The longer our telomere chains, the greater our longevity. These telomere chains gradually grow shorter with time, and when they shrink to their limit, the cell dies. Once enough critical cells die, the body itself dies.

The practices of SQ reduce the rate at which telomeres shorten. Some studies of compassion even show them growing.[22] Telomere chains are increased in length by meditation.[23]

People developing SQ are *selecting themselves for longer lifespans.* This means that the longest-lived cohort of people are going to be those who, collectively, exhibit high levels of SQ. The trend toward increasing compassion suggests that life expectancy should go up. A massive international review of health indicators in 204 countries projects an increase in lifespan of four years by 2050.[24]

Immune function is also affected by meditation. In cancer, HIV, and rheumatoid arthritis patients, it activates immune cells while improving markers of disease progress, suggesting boosted lifespan.[25]

Inflammation is often regarded as the "final common pathway in all disease" and high levels of inflammation compromise your lifespan. Research shows that in healthy patients, meditation controls inflammation and wounds heal faster. Even *three days of meditation* produces a decrease in inflammatory cells. The effect is enhanced with longer practice.

In one of my studies, we measured the effect on gene expression of four days of guided group meditation. We found significantly increased expression of eight important genes.[26] The functions of these genes are fascinating.

The first gene, DIO2, regulates metabolism by reducing insulin resistance,[27] helps reduce cravings,[28] and improves mood, especially depression.[29]

The second gene, CHAC1, helps control oxidative stress, a major cause of cell aging.[30] CTGF enhances wound repair and promotes bone and cartilage health.[31] TUFT1 stimulates stem cells, the "blank" cells from which our bodies repair other tissues.[32]

The genes C5orf66-AS1 and ALS2CL suppress cancer tumors,[33-34] while KRT24 helps build healthy cell walls.[35] The final gene, RND1, regenerates neurons.[36]

When the genes in your cells are doing all these beneficial functions, you heal quicker, reduce inflammation, repair your tissues, and live longer.

As a result of increasing SQ, expect lifespans to continue increasing. More importantly, expect health-spans to increase, with large numbers of healthy 100-year-olds running around the planet.

Expect this to have a ripple effect as they bring their talents back into society; there are only so many rounds of golf you can play before you get bored and want to do something meaningful with your life. Expect services that cater to centenarians to proliferate. Expect devices designed for the elderly to find larger markets.

Medicine

Medicine is famously averse to innovation. Its history is littered with opposition to new technologies. In the 1860s, when Ignaz Semmelweiss of Vienna advocated that doctors wash their hands before surgery, he was ridiculed and driven from his position. Though MRI scanners were invented in the early 1970s, they were largely ignored by the medical profession and it took almost two decades for scans to become widely available to patients.[37]

Today, a band of hardcore cynics control the complementary and alternative medicine (CAM) pages on Wikipedia and label it "pseudoscience." They have developed sophisticated methods that prevent the listing of peer-reviewed scientific papers on the site or updates to the CAM articles by qualified experts and reputable scientists.[38]

For instance, the first paragraph of the Acupuncture listing states: "Acupuncture is a pseudoscience; the theories and practices of TCM are not based on scientific knowledge, and it has been characterized as quackery."[39]

Go behind the Wikipedia entry to the "talk" pages to discover the opinions of the editors who control Acupuncture and the other CAM entries. One of the key gatekeepers characterizes acupuncture as "rubbish," and the National Center for Complementary and Integrative Health (NCCIH), the US Government's lead agency for scientific research on CAM, as "the pro-quackery wing of the federal government."[40] These editors prioritize their own biases over the more than 13,000 acupuncture studies published in peer-reviewed professional journals.[41] Wikipedia co-founder Larry Sanger divorced himself from the site in 2002, noting that "trolls" had taken it over and "'squat' on articles and insist on making them reflect their own specific biases" including an unremitting effort to suppress the scientific evidence for CAM.

In an important review paper I coauthored, we found *it takes 17 years for medical innovations to jump the gap from lab to patient,* and only 20% succeed in doing so.[42] That means that *just one in five effective therapies get as far as being practiced in hospitals.* The benefits of the other four are lost to society completely.[43]

Science denial such as that embodied in Wikipedia carries devastating social costs. This includes incalculable human suffering, high medical expenses, suppression of innovation, lack of patient access to evidence-based treatments, impaired wellbeing, and reduced lifespan.

Despite—or perhaps because of—this resistance to innovation, increasing numbers of people worldwide are turning to CAM treatments. This is especially true for autoimmune and lifestyle diseases for which conventional medicine has little to offer.

The nonprofit I chair, the National Institute for Integrative Healthcare (NIIH.org), maintains a public database of energy therapy studies. It is accessible by patients and caregivers at EnergyMedicineBibliography.com. It lists over 1,000 studies showing successful energy treatments for dozens of diseases from Alzheimer's to cancer to Parkinson's. Among the autoimmune diseases successfully treated are fibromyalgia, psoriasis, and lupus.

Expect the trends to continue. People with SQ are driven by curiosity more than fear. They take responsibility for their own wellbeing, desire a sense of sovereignty over their bodies, and question conventional wisdom.

Expect more highly informed patients, as they use AI to educate themselves about cures their doctors have never heard of. Expect *increased enrollment by medical professionals in CAM courses,* as responsible clinicians educate themselves about the full range of treatments.

Expect advanced labs and researchers to shorten that 17-year gap. With sophisticated imaging devices like high-resolution fMRIs and MEGs, aided by AI, we can determine the effects of SQ practices like mindfulness, tapping, and EcoMeditation in weeks rather than years. Expect pressure on governments to approve effective treatments without years of red tape.

Expect personalized medicine. For decades we've been able to take saliva swabs and urine samples and get lab results for hormones in hours. The same is now true for genes. Expect exercise, diet, and behavioral recommendations customized for each patient by doctors reading an AI-enhanced gene scan. Expect effective natural cures to soar in popularity while drugs, in many cases, become a secondary option.

Expect the stock market values attributable to conventional drugs for common lifestyle diseases to fall while the value of innovations like personalized genetic treatments rises. Expect SQ to become part of primary care, with meditation, EFT, and other natural forms of stress reduction becoming the first line of treatment.

The compassion typical of people with high SQ focus motivates them to reach out to help others who are suffering. The story of Médecins Sans Frontières (MSF) represents one shining example.

GOING BEYOND WITH MÉDECINS SANS FRONTIÈRES

Médecins Sans Frontières (MSF), or Doctors Without Borders, was founded in 1971 by a group of French doctors. The organization was born out of a desire to provide emergency medical care to populations in crisis, free from the influence of politics, religion, and national borders.

MSF's founders were horrified by the atrocities they witnessed during the Nigerian-Biafran Civil War, where the suffering of civilians went largely unnoticed by the international community. "We were faced with the obligation to speak out and intervene to help," said Bernard Kouchner, one of MSF's co-founders. MSF doctors defied the Nigerian government's restrictions and set up clandestine hospitals in Biafra, risking their lives to provide care.

From its inception, MSF has been committed to delivering medical aid where it is needed most, even in the most challenging and dangerous environments. The organization operates on the principles of impartiality, neutrality, and independence, ensuring that their medical assistance is based solely on need.

MSF's work during the Rwandan Genocide in 1994 further cemented its role as a leading humanitarian organization. As hundreds of thousands of people were slaughtered, MSF provided critical medical care to survivors, often in the midst of ongoing violence. Their presence in refugee camps, where conditions were dire and

7.5. Médecins Sans Frontières worker administering RUTF during a famine.

diseases like cholera and dysentery were rampant, saved countless lives. "In the face of such overwhelming horror, we knew we had to be there, to provide whatever care we could," recalls MSF volunteer Dr. James Orbinski.

Beyond their work in conflict zones, MSF has also pioneered medical innovations that have had a lasting impact on global health. One notable example is their use of ready-to-use therapeutic food (RUTF) to combat severe malnutrition. Introduced in the early 2000s, RUTF is a nutrient-dense paste that does not require refrigeration and can be easily administered even in remote areas.

This innovation has dramatically improved the survival rates of malnourished children, particularly in regions like sub-Saharan Africa. "RUTF has revolutionized the treatment of malnutrition," said Dr. Susan Shepherd, a pediatrician with MSF, "It allows us to reach children who would otherwise have little chance of survival."

Today MSF operates in more than 70 countries, providing medical care in some of the world's most difficult and dangerous environments. Their unwavering commitment to impartiality, independence, and medical ethics has made MSF a beacon of hope in regions ravaged by suffering and neglect. In October 1999, MSF was awarded the Nobel Peace Prize.

Psychology

Research shows that traditional talk therapy equals or beats the effects of drug treatment for many psychological conditions. Meta-analyses show that energy therapies like EFT and EMDR have even greater treatment effects.[44] These demonstrate that the field of psychology has developed the ability to cure most mental illness.

Though this is borne out by research, such treatments face an uphill struggle for acceptance—see the 17-year lag discussed previously.

Despite the odds, these treatments are so good that they will spread. With my colleagues, I wrote a comprehensive review encompassing scores of studies that was published in the top-tier journal *Frontiers in Psychology*. It showed that EFT and allied energy therapies can *remediate most cases of anxiety, depression, phobias, and PTSD.*[45] Often they require fewer than ten sessions. Sometimes just one.[46]

Practitioners certified by my organization, EFT Universe, have trained clinicians in large hospital systems like Kaiser Permanente, the Cleveland Clinic, and Johns Hopkins. In 2017, EFT was designated a "generally safe therapy" by the VA's integrative medicine department, and we now train doctors, nurses, and therapists within the VA system.

Around the same time, Korea approved EFT as a frontline treatment in primary care. There are over 100 EFT studies published in peer-reviewed journals in Chinese, Arabic, French, Farsi, Hindi, Tagalog, Japanese, Spanish, and other non-English languages, showing its global spread.

Energy therapies are often supplemented with meditation. I have conducted several studies in which EFT is combined with EcoMeditation and they show a powerful effect when used together. Meditators are naturally drawn to other "feel-good" practices.

As meditation is coupled with energy therapies, expect the number of people suffering from anxiety, depression, phobias, and PTSD to plummet. As SQ becomes widespread, expect these mental afflictions to go the way of physical diseases like smallpox, polio, and typhoid fever.

Expect a world in which mental disease is rare. As adults are cured, expect them to nurture rather than traumatize their children. Expect the mental health of future generations to be better than that of their parents.

Expect SQ to produce a *gradual downward curve in mental illness,* followed by a rapid dropoff as the older generation dies off and a psychologically healthy young generation comes online. Expect psychology to become more about unlocking human potential as its supply of treatable mental illness dries up.

Social Influence

There's nothing as powerful as an idea whose time has come. Ideas spread throughout societies, often driven by influential individuals and magnified by social media. These "influencers" are often early adopters of new ideas. Many of these are smart, well-informed people who want the best lives possible for themselves and those in their circle.

In his book *Tribe of Mentors,* uber-influencer Tim Ferriss interviews exceptional people and summarizes their power habits. He comments that though their favorite practices range the gamut, there's one commonality: meditation.[47]

SQ has pervaded public spaces since the turn of the century. Congressman Tim Ryan, elected from Ohio in 2002 when he was just 29 years old, formed the "Quiet Time Caucus" to practice meditation in the US Capitol. Medtronic founder Earl Bakken set up a meditation room at the company headquarters in 1974. Goldman Sachs offers its employees meditation classes. Basketball player LeBron James has been a regular meditator for years, famously meditating during a time-out in the 2012 NBA playoffs.

7.6. A team of athletes meditating.

After the Seattle Seahawks won the 2014 Super Bowl, their coach, Pete Carroll, told reporters he'd hired psychologists to teach meditation to his athletes and get them into flow states.

Pop singer Katy Perry told a reporter, "I start the day with Transcendental Meditation. It puts me in the best mood."[48] In the 1960s, the Beatles studied meditation with Maharishi Mahesh Yogi.

Band member Paul McCartney said, "In moments of madness, meditation has helped me find moments of serenity."[49] Many other celebrities meditate, including Beyoncé, Hugh Jackman, Lindsey Lohan, Santigold, Amy Schumer, Jerry Seinfeld, Clint Eastwood, and Madonna.

Expect the adoption of meditation by influencers to continue and for their advocacy of SQ practices to increase as they experience its benefits in their lives and the lives of those around them.

Creativity

Expect creativity to continue its trend of acceleration in all fields. One of the defining characteristics of high-performing executives, according to the McKinsey report referenced in Chapter 5, is the ability to *hold a variety of conflicting ideas in mind at the same time.*

For many years I managed a book publishing and distribution company. One day our sales manager came to my office to talk about a crisis. He said, "We can either do this (unpleasant option A) or we can do this (unpleasant option B)."

"Come back to me when you have at least five options," I replied.

I knew that while his mind was stuck in either/or thinking, perceiving two options, he was trapped in a limited conceptual box. Encouraging him to imagine multiple possibilities got his creative juices flowing.

People with high SQ enjoy playing in their heads with all kinds of different possibilities. The McKinsey study found that many of these may be mutually exclusive. But allowing them space opens up the range of creative options, and in the flow state, surprising new syntheses may be found.

This explosion of creativity will leave no field of human endeavor untouched. The 21st century has witnessed the disruption of even stodgy old industries like logistics, mattress manufacture, home appliances, bicycle design, and winemaking. Expect to be startled, amazed, and delighted as you see innovation flourishing everywhere.

Technology

Technology is already one of the most innovative fields of human endeavor. Gadget manufacturers study the advances made by their competitors, reverse engineer them, and improve on them. This results in a rapid product development cycle.

Those with SQ are mindful about their use of technology. Because they aren't fearful, they aren't reactive to the unscientific fearmongering that often surrounds technology (Remember the Y2K bug? Malevolent computers taking over the earth like HAL in *2001: A Space Odyssey*? Killer robots like in the movie *Terminator*? Microchips in covid vaccines? AI eliminating human jobs?) A poll showed that Americans are more afraid of robots than death.[50] But those exemplifying SQ are curious and open.

They don't embrace every new technology either. They prize time in nature, emotionally intimate connections with other people, self-reflection, exquisite experiences, and other organic activities above technologically mediated ones. Yet they're excited by the potential of technology and enjoy it when they use it.

The creativity unleashed by SQ will affect the development of new technologies too. Expect innovation to accelerate at a pace even greater than today, but also expect it to *be inclusive of social good.*

Expect technological innovation to be pushed less by military need—the primary driver in the 20th century—than by the demands of human flourishing. Expect the flood of creativity unleashed by SQ to produce entirely new technologies not even dreamed of today.

Science

Expect calm brains to produce continuous breakthroughs in every domain of science, from mathematics to physics to chemistry to geology. When your attention isn't hijacked by your emotions, you have more intellectual capacity available for discovery.

You also have greater access to your intuition. Albert Einstein claimed that intuition was the source of every great scientific innovation. His theory of relativity came to him in a dream, after many months of frustration. He then spent several years working out the mathematics.

Increasing numbers of scientists are meditating, joining the ranks of the SQers. As we saw in Chapter 2, this is associated with anomalous experiences like clairvoyance, telepathy, synchronicity, and precognition. Expect the 200-year-old divide between science and spirituality to crumble.

As more scientists develop SQ, expect their happiness levels to increase, their creativity to explode, and their output of innovations to soar. Though the scientific output of the past century seems astonishing, we're just getting started. Expect the breakthroughs produced by science to revolutionize every other domain of human activity in the coming century.

War and Violence

War is already obsolete; expect it to become unthinkable.

In its 1949 constitution, Costa Rica became the first country in the world to abolish its army. This spared it the horrors of the civil wars that wracked its Central American neighbors, Guatemala, Honduras, Nicaragua, and El Salvador. In those countries, military juntas alternated with civilian rule, often seizing power from elected officials. At least 300,000 citizens of those four countries died and over two million were displaced.

That wasn't possible in Costa Rica because there was no army to take power. Starting with a level playing field in 1949, the country's economic development soon outstripped its neighbors. Today the average Costa Rican enjoys between *twice and five times the income* of citizens of those neighboring countries. The country invested its peace dividend in education and healthcare.

Compassionate brains steeped in SQ don't pick fights with others and respond wisely when attacked. For the last 500 years, nations have become progressively more reluctant to engage in all-out war with each other. This shows up in the data on regional and small-scale wars, and especially in the case of great power wars. Great powers are nations that are capable of projecting force beyond their borders.

Go back to the 1500s and we find that the great powers were almost always at war with each other. The names of some of these wars convey a sense of their duration: the Hundred Years War between England and

France (1300s), the Eighty Years War for Dutch Independence (1500s), and the Thirty Years War between shifting coalitions of European states (1600s).

7.7. Wars between Great Powers from 1500 on.[51]

However, as can be seen from the above infographic, wars between great powers have been declining for the past 500 years. While today's headlines highlight regular armed conflicts, historical data show that *human beings have been getting collectively more peaceful for centuries.*

Countries other than Costa Rica have disbanded their armies. When Panama followed suit, it eliminated the threat of the military coups that had plagued its history. Some 20 other countries including Iceland, Dominica, Mauritius, and Andorra have also abolished their armed forces, reaping social and financial dividends.

Smart brains, steeped in SQ, find alternatives to war. They are able to avert potential conflicts and imagine new solutions that minimize the possibility of war. They focus on how to create peace and take concrete steps to produce it.

There are catalogs of wars and conflicts, yet what draws less attention is lists of "Peace Events" that enumerate the practical steps human communities are taking to "give peace a chance." Here's a list of "Peace Events" from just a single decade, 1901 to 1910:

December 10, 1901: First Nobel Peace Prize awarded.

January 16, 1906: Algeciras Conference produces a framework to regulate the trade in weapons.

July 6, 1906: Signing of the first Geneva Convention regulating treatment of enemy combatants. The second was signed in 1929.

September 11, 1906: In Johannesburg, South Africa, Mohandas Gandhi launches a peaceful protest campaign against racism.

June 15, 1907: The Second Hague Conference provides for arbitration of international disputes instead of war.

November 13, 1908: Mohandas Gandhi publishes the principles for nonviolence as the means for political change.

November 25, 1910: The Carnegie Endowment for International Peace is established by Andrew Carnegie to "hasten the abolition of international war, the foulest blot upon our civilization."

These events are markers of a level of SQ that would have been inconceivable a century earlier. For dozens of other inspiring examples, organized by decade, see the Extended Play section at the end of this chapter.

Look particularly at the list from the 1960s. During that decade, US President John F. Kennedy created the Peace Corps; Amnesty International, an organization dedicated to freeing political prisoners, was founded; the Cuban missile crisis averted a nuclear missile exchange between the US and USSR; the Organization of African Unity was established to resolve African conflicts; the treaty demilitarizing outer space was negotiated; and the Nuclear Non-Proliferation Treaty was signed.[52]

All of these required both SQ and an immense effort by inspired people to change norms of war and conflict that had been around since the dawn of human history. One of the most inspiring examples of wisdom during war comes from the transcript of the trial of Joan of Arc, patron saint of France.

In God's Grace

Heavy cold iron chains dragging her wrists toward the floor, the slender young woman faced her accusers. Just 19 years old, she had been held in a military prison for a year, interrogated nearly a dozen times by a tribunal, and threatened daily with rape and torture. Her name was Joan of Arc and the year was 1431. Two years earlier, just 17 years old, she had led the French army to victory at the Battle of Orleans.

The battle was part of the "Hundred Years War" between France and England. During this period, towns and villages on the border between the two countries were racked by uncontrolled violence. Organized government broke down, leading to lawlessness and frequent attacks on civilians. Bands of brigands called *routiers* raped, looted, and pillaged at will, leading to floods of refugees fleeing these areas. Until a strong French king emerged, the chaos would continue.

Raised by a pious mother, Joan began having mystical visions when she was just a child. She dreamed that Saint Michael and Saint Catherine were inspiring her to support the claim of Charles VII to the French throne. Joan's aim was a strong kingdom that could repel the English invaders and bring peace to France.

In 1429, supporters of Charles's claim sent her to his court. That required an 11-day journey across enemy territory. Joan cropped her hair and dressed in men's clothes to minimize her conspicuous presence among an escort of soldiers. Anticipating her arrival, Charles had no idea what to do with this peasant girl who claimed she could save France from its English and Burgundian enemies.

Charles put her to a test. He dressed incognito and mingled with members of his court. Joan, who had no idea what he looked like, identified him immediately. After meeting with Joan, Charles had her examined by prominent theologians. They found only humility, chastity, and piety in the girl.

Charles gave her a horse and armor and she accompanied the army to the city of Orleans, which was besieged by the English. In a four-day battle, the French gained a foothold in the English fortifications, though Joan was wounded.

7.8. 1903 Depiction of Joan of Arc by Albert Lynch.

By the middle of the following month, Joan returned to the frontlines to participate in a final assault. Inspired by the "maid of Orleans," the French routed the English army. When Charles was crowned king on July 18, 1429, Joan was at his side.

The following spring, the king sent Joan to the town of Compiegne to repel a fresh attack by the Burgundians. She was captured and turned over to church officials. They tried her as a heretic, charging her with 70 counts, including heresy, witchcraft, and wearing male garb.

Rather than being imprisoned in a church with nuns guarding her, as was the custom, Joan was held as an ordinary captive in a military prison. She tied her body tightly with cords to protect herself against rape.

On May 29, 1431, the tribunal announced its decision. It found Joan guilty of heresy. On the following morning, she was burned at the stake in the center of Rouen in front of 10,000 people.

Twenty-five years later, an investigation ordered by the king found Joan innocent of all charges. In 1920, the Roman Catholic Church canonized her as a saint.

During Joan's trial, her cunning captors devised many trick questions designed to trip her into a confession of heresy. One of these was "Are you or are you not in God's grace?"

They knew Joan would be damned whichever way she answered. If she said "No," indicating she was not in God's grace, then she would have confessed to being guilty of apostasy. If she said "Yes," indicating she was in God's grace, her accusers would be able to say she had claimed knowledge of the mind of God, an equally serious heresy.

To this impossible question, Joan confounded her accusers by replying, "If I am not, may God place me there; if I am, may God so keep me."[53]

When people with SQ are asked impossible questions, as Joan was, they find wise solutions that transcend either-or thinking. SQ gives us the wisdom to get out of either-or thinking and boxes that limit the range of possibilities.

As humankind has become progressively more compassionate over the past millennium, the effect shows up in metrics other than the communal violence of war. It also shows up as a reduction in the personal violence of murder.

Data show that the homicide rate has been declining over the past 800 years. Collectively, the compassionate people of today are much less likely to resort to killing their neighbors than their grandparents were, and documentation for this trend goes back almost a thousand years.

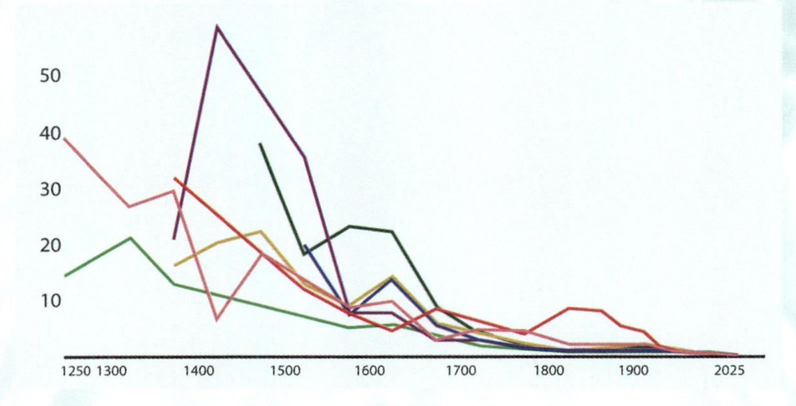

7.9. Number of homicides per 100,000 people in the European countries from 1250 to the present.[54]

Expect the trend of decreasing personal and communal violence to continue as SQ spreads. While wars may flare up and interpersonal violence might not disappear, the trends apparent in data from the last millennium suggest that both are likely to diminish in the coming decades.

Politics and Government

People with SQ are inclusive. They don't see the world as us versus them, and they don't vote for people who do. Happiness leads naturally to tolerance, respect for human differences, and enlightened policy.

As SQ gains ground globally, expect politics that appeals to fear to lose traction. Expect win-win arguments to supersede winner-takes-all thinking. Expect the politics of the common good, expanding to include the good of all the people in the world, to grow in appeal. Expect nationalism—and in-group versus out-group rivalries—to wane. Expect integrity to invade politics, however unlikely that may seem today.

The values of SQ, such as compassion for other people, show up in public policy. In developed countries, the proportion of government spending on social projects such as education and relief for the poor totaled 2% at the start of the 20th century. Today, the average is 22%.[55]

That's an 11-fold increase. Likewise, the global measures of human well-being graphed at the start of this chapter show that education, healthcare, security, and similar metrics are improving in all regions of the world.

Expect equitable social policies to grow in appeal and be voted in by the very people who will have to pay for them. Expect corruption to decline. Those with SQ have shrunk the neural pathways of fear and greed to the point where they lack the hardware to contemplate the exploitation of others.

Environmental Activism

SQ leads to caring about other people and the planet. Meditators behave proactively when it comes to nurturing the environment. Environmental "awareness" or "stewardship" is not enough. We need to be proactive to undo the damage done to the planet by past generations and the overwhelming numbers of present-day humans.

Spiritually intelligent people will apply their vastly increased creativity and problem-solving ability to the wicked problems of today. Mass extinction. Climate change. Deforestation. Rising sea levels. Groundwater depletion. Space junk. War. Habitat loss. Income inequality. Dirty fuels. Industrial pollution. Gender issues. Agribusiness. Declining crop yields. Science denial. Topsoil erosion. Short-term thinking.

Expect many of these problems to be solved. Some will be solved quickly and in surprisingly simple ways. Others will take effort and expenditure at the level of nation-state cooperation. Either way, expect the mental resources of SQ propagated through collective culture to be able to resolve environmental issues that are insoluble by Caveman Brain.

SQ perceives the planet Earth as an organic whole, with humans as one component of the mix. This mental frame promotes holistic, long-term solutions. Expect environmentally active policies and mindsets to become the new paradigm. We might also have a lot of fun along the way, as the story of EVs in Norway illustrates.

CRAZY FUN VISIONARIES

In the mid-80s, Norway's electric vehicle (EV) revolution was ignited by an unlikely trio: an activist, a professor, and the lead singer of the band A-ha.

Architect Harald N. Røstvik, disillusioned by Norway's rush into oil production in the North Sea, teamed up with environmentalist Frederic Hauge and A-ha's Morten Harket. Together, they purchased a makeshift electric car—a converted Fiat Panda—and hit the roads of Norway with a mission to challenge the government's policies on EVs.

The car looked like a joke. The back seat was ripped out to accommodate a bank of lead-acid batteries. It took two days to charge, after which it had a range of just 40 miles.

7.10. The mad electric crew.

But with Harket's star power leading the way, the group made a splash. They flouted the rules of the road, driving in bus lanes, skipping tolls, and parking wherever they pleased. Their antics were not just rebellious but strategic, aiming to draw media attention to the need for tax breaks and incentives for EVs. "Because we had Morten and Mags, we got enormous attention," Hauge recalls. Their defiance paid off, forcing the government to rethink its stance on electric vehicles.[56]

In the late 90s, Norway was one of the first countries to introduce incentives like free parking, toll exemptions, and bus lane access for EVs. These policies laid the foundation for the country's current EV dominance. Today, around 80% of new cars sold in Norway are electric, with the country aiming for a fully electric car fleet in the near future.

To Røstvik, it all began with a change of consciousness. "Back then," he observes, "there was laughter all around. Electric vehicles, it will never be realistic; solar energy, never; wind power a joke. And look where we are now. Everything has changed."

Agriculture and Farming

Respect for other life forms is a cornerstone of SQ. In 2008, California voters passed an initiative by 60% that required more humane treatment of egg-laying hens. While opposed by the industry and ridiculed by opponents, it became the new normal. Today the movement for the humane treatment of animals is being extended to other species such as pigs, cows, and farmed fish.

As compassion and empathy take hold, many SQers become vegetarians, and even if they don't, they believe that animals have the right to decent treatment.

At the same time, the technology for growing meat in the lab has matured. Large-scale cultivation is underway, backed by billion-dollar stock IPOs and massive advertising campaigns. Some of these "meats" are manufactured from yeasts and other plant cells; others from animal muscle cells cultivated in nutrient solutions.

Expect consumption of these meat alternatives to overtake that of meat from live animals in a few years. Expect the latter, in a few decades, to be an exotic rarity reserved for gourmets and deviants.

People with SQ care about what they put into their bodies. The demand for organic produce has exploded in the past few years. The percentage of US cropland designated as "organic" doubled between 1990 and 2002. It doubled again by 2005. A decade later, it had expanded by nearly 80%. Organic animal-raising operations have grown even faster, according to the US Department of Agriculture.[57]

Expect an increase in the number of vegetarians. Expect the trend to organics to accelerate. Expect people to demand accurate labeling of foods. Expect increased demand for high-quality food. Expect the compassion embodied by SQ to result in the more humane treatment of all animals. Expect the stock market values of fertilizer and pesticide companies to drop, and those of natural farming businesses to rise.

Climate Change

Climate change is one of the most serious problems faced by our species and the planet. The disruption to all areas of human life promises to be dramatic and exponential unless we implement solutions quickly. What's often forgotten when looking at the problems in front of us, however, is the solutions behind us.

Remember the ozone layer? It is a protective shield in Earth's stratosphere that absorbs most of the sun's harmful ultraviolet (UV) radiation. By the 1980s, scientists had measured a big hole in the ozone layer over Antarctica that was allowing copious amounts of UV radiation to reach Earth's surface, increasing skin cancer, cataracts, and disruptions in the air flow around the planet.

The world's governments promptly got together and banned the chemicals that were producing the hole. In a remarkable environmental success story, ozone depletion was reversed, the hole shrunk, and climate disruption attenuated. That's what human beings can do when smart and compassionate brains combine with concerted social action.

We've now developed technologies to pull carbon and methane from the air. This isn't a sci-fi fantasy; there are currently dozens of carbon removal factories operating worldwide.[58]

Another existing technology mimics the effect of volcanic eruptions. In 1991, Mount Pinatubo in the Philippines erupted, spewing volcanic ash into the air. This reflective aerosol of suspended particles reduced the amount of sunlight reaching Earth's surface.

This in turn cooled the planet's entire atmosphere by about 1° Fahrenheit or 0.5° Centigrade. It made 1992 and 1993 the coolest years in the last half of the 20th century. International scientists have figured out how to mimic this natural process, enabling us to reverse global warming swiftly should we muster the collective will to do so.[59]

The manufacture of concrete is responsible for 8% of global greenhouse gas emissions. New technologies cut this by 90%.[60] While deforestation is a global problem, the world's two most populous countries, China and India, have significantly reforested since the turn of the century.[61]

Some countries have done even better. Norway has *tripled* its forest cover in the past century.[62] The Earth's green area has increased by 5% since the early 2000s, according to NASA satellite surveys. That area is so large that *it's the equivalent of adding a second Amazon rainforest* to the planet.[63]

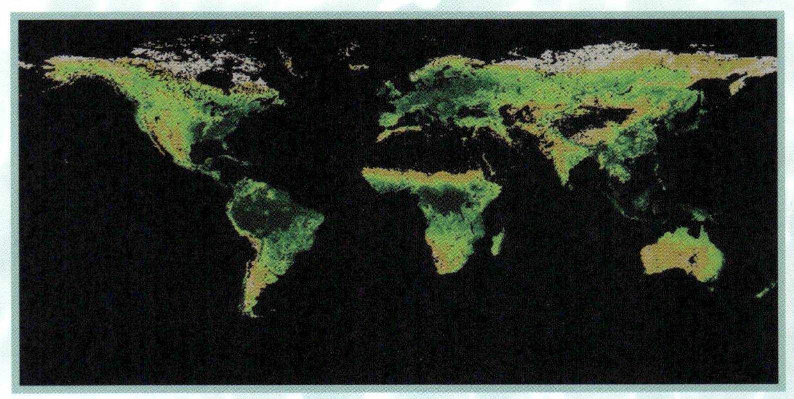

7.11. NASA map showing substantial reforestation since 2000.
The tan color indicates little or no leaf cover.[64] Light green indicates one layer of leaves.Dark green shows thick forest canopies with seven or more layers of leaves. Black means no data.

Addressing climate change isn't just good for the environment; it's good for the economy. *The costs of mitigating climate change are actually lower than doing nothing.*[65] Expect smart and creative brains steeped in SQ to find new solutions and scale up the implementation of existing ones.

Religion and Philosophy

As we saw at the start of Chapter 3, research shows that people who have mystical experiences become simultaneously less religious and more spiritual. Most religions are actually shrinking in size.

That doesn't mean that religion is going away. Instead, expect religions to converge in the mystical experience from which they all sprang. Today the world's largest single designation is "spiritual but not religious."

Expect the trend to continue, but also include reverence for the religious traditions that kept SQ alive during epochs of cultural darkness. Expect philosophy to increase its focus on compassion, gratitude, and the other positive emotions of SQ.

THE POPE AND THE ASSASSIN

On May 13, 1981, St. Peter's Square in Rome, center of the Vatican and the Catholic Church, was filled with the faithful, eager to catch a glimpse of Pope John Paul II during his weekly audience.

As the pope moved through the crowd, smiling and blessing those around him, the air suddenly erupted with the sound of gunfire. Mehmet Ali Ağca, a Turkish gunman, fired several shots at the pope from close range.

Chaos erupted as the pope slumped into the arms of his attendants, blood streaming from his white cassock. The bullets had struck his abdomen, causing massive internal bleeding. The situation was dire; the pope was rapidly losing blood.

7.12. Blood on his hand, Pope John Paul II is assisted by aides moments after he was shot.

He was rushed to Gemelli Hospital. Doctors fought to save his life in a lengthy and grueling surgery, unsure if he would survive. In what many regarded as a miracle, Pope John Paul II pulled through.

Two years after the shooting, the pope visited Ağca in prison, where the two men spoke privately. The pope held Ağca's hand, offering him words of forgiveness and compassion. This supreme act of SQ didn't end there; the pope later petitioned for Ağca's pardon, demonstrating his deep commitment to forgiveness and love, even toward those who had caused him great harm.

Pope John Paul II's actions transcended human instinct for retribution, showcasing the power of SQ to foster peace and understanding in the face of violence.

Mapping the Void

While the language used in traditional descriptions of mystical experiences varies greatly from tradition to tradition, both Eastern and Western mystics use a common concept for the experience of universal consciousness: the Void.

Paul Bruton called it the Great Silence. Kabbalah uses the term "mystical nothingness," while Buddhism emphasizes "emptiness." Meister Eckhart refers to "changeless existence" and "nothingness beyond being." Anandamayi Ma spoke of the dissolution of individual consciousness as it enters *mahasunya* or the Void.

What is the nature of the Void of which the mystics speak? Some describe it as being anything but empty, as pulsing with divine light. Others say it is filled with endless energy. Yet others describe vast information fields populated by infinite intelligence.

Expect science, philosophy, and spirituality to collect the experiences of travelers who've been to the realm of pure consciousness itself. Academics of the future will study *the nature of undifferentiated consciousness* by analyzing the experience of finite human minds that have entered a state of communion with it.

A century ago, quantum fields and electromagnetism were still mysterious phenomena about which we had limited knowledge. Science has now pieced together coherent pictures of them. Expect new branches of science and academia to map the Void in the same way.

Management

In Chapter 5, we saw how SQ is a game changer for organizations. My "feet on the ground" study showed that it boosted productivity by 20% in just 30 days.[66] Other research cited in that chapter shows that SQ produces better financial performance, job satisfaction, workplace performance, life satisfaction, resilience, leadership effectiveness, and integration of brain function.

In the book *The Art of Fairness: The Power of Decency in a World Turned Mean,* there are many examples of businesses that succeed by being nice.[67] One of them describes the construction of the Empire State Building. From its completion in 1931, it held the record as the world's tallest building for 40 years.

It was erected in just one year and one month, a time frame that also included the demolition of a Waldorf-Astoria hotel that previously occupied the site.

The builder, Paul Starrett, showed a high degree of concern for the wellbeing of his construction workers. He paid them even on days when high winds halted work. Daily wages were more than twice the typical rate and hot meals were provided on-site. He was meticulous about worker safety.

7.13. Lunch 800 feet in the air.

Starrett's approach echoes research demonstrating that companies that pay workers well and treat them fairly tend to attract more skilled and motivated employees. The Empire State Building project saw much lower staff turnover than normal. Workers were so enthusiastic about the success of the scheme that they suggested ways to boost productivity, such as installing a miniature railway to transport bricks around the site quickly.

However, Starrett wasn't blindly generous. He employed accountants to monitor the worksite to ensure that all materials were accounted for. He recorded worker attendance four times each day. His balance of good management and fair treatment of people modeled enlightened leadership.[68]

In the future, expect organizations to encourage SQ for both intrinsic and financial reasons. It makes workplaces more pleasant and attractive, as well as contributes to the bottom line.

Economics and Fiscal Policy

As the health advantages of SQ spread through societies, expect changes in the way healthcare dollars are spent. Expect costs for meditators to shrink, though the costs for those relying on conventional medicine may continue to rise. Expect voters to question the amounts governments are spending on "defense" (aka military might) as opposed to social good.

Awakeners I know, especially the rich ones, have scant tolerance for huge gaps between rich and poor. Expect creative approaches to restructuring economies to work for everyone. Expect ingenious incentives for both rich and poor to help bring the world's wealth closer to balance.

Expect short-term policies—like running up huge government debt, spending deficits, and unfunded liabilities—to fall out of favor. The calm and creative minds of SQers are capable of doing the math. Expect them to find ingenious ways out of these dilemmas and to reward leaders who participate in discovering solutions.

Wealth

Decisions about wealth—acquiring, investing, and spending it—are driven by values. What we believe shapes the way we perceive and act around wealth, whether that wealth is personal or societal. People with high levels of SQ have values, like altruism, human rights, sustainability, compassion, and fairness, that drive their approach to wealth.

In recent years, these values have produced a significant shift in the world of investing. This new approach, called "values-based investing" or "impact investing," contrasts sharply with the traditional focus on maximizing financial returns at any cost.

Today's investors examine the *ethical implications of their investments,* supporting companies and projects that align with their moral principles. This trend reflects the broader cultural movement toward social responsibility and sustainability, marking a departure from the profit-driven mindset of previous generations.

One vivid example of how values-based investing differs from traditional approaches can be seen in the rise of ESG (Environmental, Social, and Governance) funds. Unlike conventional funds that might invest in any profitable company regardless of its impact, ESG funds actively exclude businesses involved in harmful practices, such as fossil fuel extraction, tobacco, or labor exploitation. Instead, they channel capital toward companies that promote renewable energy, fair labor practices, and corporate transparency.

The very first ESG fund, the Pax World Fund, was formed in 1971. It was created by two United Methodist ministers, Luther Tyson and Jack Corbett, as an investment option that avoided companies involved in the Vietnam War. It pioneered the ESG concept and laid the groundwork for investors striving to create a positive impact on society and the planet with their wealth.

Wealthy people are increasingly rethinking their approach to passing on their wealth as well. Rather than simply enriching their children, they often choose to invest in a better world. Instead of leaving vast fortunes to their heirs, many are directing their estates toward *philanthropic causes that address pressing global issues such as climate change, poverty, and social inequality.*

This shift reflects the recognition inherent in SQ that with wealth comes power and responsibility. Awakeners prioritize the wellbeing of humanity and the planet over the accumulation of personal or family wealth. They *align their wealth with their values,* funding initiatives that create lasting positive change and leave behind a legacy of compassion and progress.

THE GIVING PLEDGE

Ted Turner and Warren Buffett have played pivotal roles in reshaping the approach to wealth distribution among the ultra-rich. Ted Turner is an American media mogul and philanthropist, best known for founding Cable News Network (CNN). Warren Buffett, aka the

"Oracle of Omaha," is one of the most successful investors of all time. His conglomerate Berkshire Hathaway has outperformed the stock market average for half a century.

In 1997, in a move that stunned the philanthropic world, Ted Turner donated $1 billion to the United Nations. He called unapologetically for changing the definition of success to wealth distributed for the social good, not wealth *accumulated for personal gain.*

Warren Buffett further revolutionized this mindset in 2006 by pledging to give away the majority of his fortune, inspiring others to do the same. This radical departure from the traditional practice of passing wealth down through generations culminated in the creation of the Giving Pledge in 2010.

Created by Buffett and Bill Gates, cofounder of Microsoft, the Giving Pledge encourages billionaires to commit at least half of their wealth to charitable causes. This movement has since garnered support from an increasing number of wealthy people worldwide, marking a significant cultural shift toward using wealth to address global challenges rather than enriching one's descendants.

7.14. Hundreds of the world's wealthiest have committed to the Giving Pledge.

Over 240 billionaires from around the world have signed the Giving Pledge, including Richard Branson, Melinda French Gates, Michael Bloomberg, Mark Zuckerberg, and Sara Blakely. It marks a complete shift in the approach of the ultra-wealthy in the direction of global responsibility.

For those of more modest means, there's the Giving What We Can (GWWC) pledge. These are people who pledge to donate at least 10% of their income to charity. Research shows that doing good improves your wellbeing significantly. GWWC pledgers have higher levels of prosocial behavior, open-mindedness to new ideas, more complex decision-making processes, and greater empathy and compassion.[69]

Professions

Half the job descriptions of today did not exist 50 years ago. Blockchain developers. Podcast producers. AI ethicists. Cryptocurrency analysts. Virtual physicians. Cybersecurity data scientists. Web developers. VR engineers. Molecular biologists. Nanotechnologists. Gene technicians. Drone operators. Bioethicists. Self-flying car designers. Nurse navigators. Smartphone repairpeople. The list is endless.

Fifty years from now, in a more peaceful and compassionate world, we'll need fewer soldiers, policemen, judges, lawyers, and prison guards. With a healthier population, we'll need fewer nurses, psychotherapists, and doctors. We'll enjoy the services of more creatives, innovators, and educators.

SQers will follow their interests. Their active minds won't tolerate a long career in a boring job just because there's a pension at the end. Expect them to produce disruptive innovation, even in traditional industries.

Expect them to invent entirely new occupations that nurture the human spirit while fostering unbridled creativity. Expect the lines between work and play to blur. Expect a raft of new professions and job descriptions that describe the construction of a happier and more compassionate world.

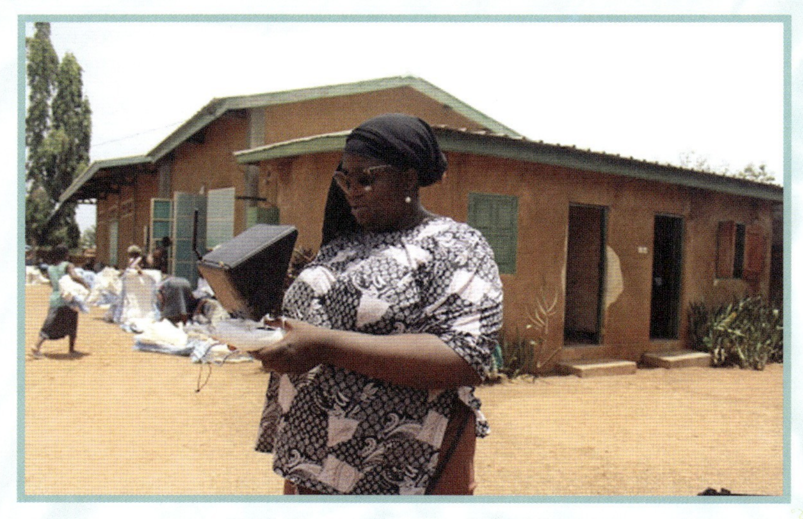

7.15. Drone operator in Africa.

Business

There's a memorable scene in the movie *The Founder,* a biopic about Ray Kroc, builder of the McDonald's hamburger franchise. Referring to competition, he says, "It's dog eat dog, rat eat rat. If my competitor were drowning, I'd walk over and put a hose right in his mouth."

Such attitudes will become increasingly rare in a spiritually intelligent world.

New models of business are evolving. One is affiliate partnerships. Groups of businesses with similar customer bases used to perceive each other as competitors. In affiliate relationships, they support each other and they all grow. In this new model, competition is not the key to success. What is?

Cooperation.

It's natural in people with well-developed SQ. In Chapter 5, I described inducing group flow in work teams. For a few minutes before a team meeting, I lead members through a routine that creates heart coherence and a stable alpha state. We do this before any discussion of the agenda occurs because creating the right mindset is the source of productivity.

Only after group flow is induced does the meeting begin.

I've even done this among warring tribes. Not the kind covered in finger paint and wearing loincloths, but competing business organizations. When they come together in group flow, with an appreciation of each other's humanity, disputes can be quickly resolved and common ground identified.

7.16. SQ improves productivity and mood in business.

One of my studies involved members of the World President's Association, entrepreneurs who had built businesses with an annual income of $9 million or more. We found that *in a single day* of using stress-reduction methods like EFT tapping, their levels of anxiety and depression *dropped by 34%.* Pain was reduced by 41%, and cravings for problem food and drink items by 50%.[70] Imagine people in millions of businesses getting access to the game-changing results produced by EFT and meditation. The studies described at the end of Chapter 5 show that *productivity skyrockets as stress falls.*

Expect these practices to become routine in spiritually intelligent businesses. Expect high-quality human relationships to become a core value of the business enterprise.

Expect models of competition to give way to those of collaboration and co-creation. Expect business to emphasize social good as well as profit. Expect win-win transactions to become axiomatic. Expect this to unleash a wave of productivity before which the gains of the past century pale. Expect this bounty of socially responsible and environmentally sustainable wealth-creation to improve the lives of billions of people.

Law

We'll always need lawyers to draw up wills and property transfer deeds, though AI and VR will increase automation of these tasks. But in an enlightened society, we're less likely to see plaintiffs exploited in the service of pointless litigation designed only to enrich lawyers.

One example is no-fault auto insurance. Many US states use this method of setting the liability for car crashes. Each insurance company covers the cost of its own driver, completely avoiding the need for legal proceedings.

When a 1988 ballot initiative proposed this reform in California, trial lawyers saw a rich source of income going away and poured money into an advertising campaign to defeat the reform. At a cost of $60 million, it became the costliest election race in California history except for the presidential election.

Those whose minds are filled with SQ aren't willing to sacrifice their peace of mind by going to court. Expect disputing parties to use mediation, arbitration, and other ways of settling their differences amicably.

Expect many laws to vanish or become unnecessary.[71-72] The North Carolina legislature has decreed that bingo games can't go on for over five hours. Utah says that cousins can't marry if they're under 65 years old. Under the state's indecency laws, Montana lawmakers took up a proposed law banning yoga pants. Canada bans the sale of hamburgers cooked any less than well done. Michigan state's senate approved a law making oral sex a felony, even if it's performed by married heterosexual couples. Can we live without laws like this? Certainly.

Expect crime rates to drop. Expect prison populations to fall and many prisons to close. Some will become museums, others given a heavy dose of feng shui then repurposed for useful ends. Expect laws designed to regulate bad behavior becoming obsolete because there's less harmful behavior left to regulate. Expect lawyers to be working on fewer disputes and more creative collaborations between parties.

Parenting, Marriage, and Intergenerational Relationships

Unconscious parents often inflict the wounds of their own past on their offspring. As parents become more self-aware, expect many to clean up their own psychological damage before they have children. *Expect parents to be less likely to project the deficits of their own upbringing* onto their children.

I wrote a summary of the latest research into PTSD. It's a 100-page booklet entitled *Psychological Trauma: Healing Its Roots in Brain, Body, and Memory.*[73] In it, I review the studies on the effects of child abuse and neglect. Those effects are biological as well as psychological.

Abused children have abnormal hormonal profiles, with both sex and stress hormones dysregulated. They have high rates of cancer, diabetes, and depression.

Expect the amygdalas of billions of parents to atrophy, while their Enlightenment Networks grow. Expect these peaceful happy parents to raise even more peaceful and happy children. *Expect this process to repeat itself across generations, evolving the brain anatomy of our species progressively.*

7.17. Compassion breeds connection across the generations.

Speaking from my own experience, the intense pleasure that results from a deep spiritual and emotional bond with your children has few parallels. Expect the experience of parenting to change for both parents and children. The joy of connection will be celebrated and sources of conflict diminished.

Expect this to enhance not just childhood bonding but lifetime bonding between generations. Expect compassionate multigenerational relationships, with increased support and collaboration up and down the timeline.

Expect marriages to become more harmonious, as partners employ the tools and values of SQ to build high-quality relationships.

Education

Meditating parents are calm parents. They provide an emotionally and spiritually richer environment for their children. They want the same values in the schools their children attend. SQ will produce wiser and more understanding teachers, and foster caring and support among students.

Expect the children of spiritually intelligent parents to carry the habits of SQ to school. Expect bullying, social anxiety, test anxiety, and similar stresses to diminish. Expect classrooms to become places where love, peace, and harmony are commonplace. Expect the learning curve of children in these nurturing environments to improve. In this supportive setting, expect IQ and test scores to increase.

Art, Literature, and Architecture

Expect the creativity unleashed by SQ to permeate all forms of art, literature, and architecture. This might lead to entirely new and original art forms.

Expect nature, spirituality, nurturing, and harmony with the planet to be incorporated into new architectural designs. Buildings will increasingly emphasize meditation spaces, green elements, mindfulness cues, sustainability, and environmentally responsible materials.

7.18. Artist Pablo Picasso creating artwork by waving a light.

Expect the values of SQ to be expressed in the arts. Expect a decrease in music lyrics celebrating misogyny, violence, and negativity. Expect the consumption of soothing and stress-reducing artistic expression to increase.

Journalism

"If it bleeds, it leads" has been the motto of news organizations for decades. The study cited at the start of this chapter measured the increase in negative words and decrease in positive ones over the past century. This has taken journalism further and further away from the thriving that is the global norm.

Expect the *proliferation of positive news* and outlets to broadcast it. There are many of these already. Personally, after I've looked at conventional news platforms, I check out sites like Nice.news, Positive.news, and the Good News Network to balance out my diet of information. Expect mainstream journalism to begin reporting more good news stories as SQ produces the huge shifts in human experience and behavior envisioned in this chapter.

Sports

Top athletes understand that peak performance occurs when they are in a state of flow. Expect methods of inducing flow states, like EFT tapping and EcoMeditation, to become a standard part of training. A grasp of the body's energy flows might become as useful to athletes as understanding muscle groups.

Athletes will always compete with each other, but the competitions are unlikely to become grudge matches. Appreciative people enjoy the peak performance of another person, even when bested.

Expect to see records broken regularly. SQ produces more efficient brain function, which in turn leads to performance improvements. Expect athletes to compete against their own "personal best" standards as much as against others. Expect violent sports to decline in popularity, while games of skill endure.

Expect to see teams playing together more cohesively. I've worked with sports teams as well as business teams, and when they're in group flow, their performance multiplies. Expect coaches everywhere to teach the value of these methods.

THE BEAVERS TURN ON THE EMOTION REGULATION CIRCUIT

Greg Warburton, a therapist with a passion for sports performance, has always been immersed in athletics. After completing his first marathon at 27, a tragic motorcycle accident cost him a leg. Undeterred, Greg pursued a career as a therapist, where he specialized in teaching high-performance techniques like EFT, meditation, and focusing to athletes.

In a groundbreaking study at Oregon State University (OSU), Greg and I collaborated on a randomized controlled trial with the men's and women's basketball teams, the OSU Beavers. We measured each member's free-throw accuracy and jumping ability.[74]

One group then received a brief EFT session, while the other group was given a placebo treatment. After just 15 minutes of EFT, the athletes improved their free-throw performance by 38% compared to the control group.

Observing the players on the court, I noticed a striking pattern. Some players were consistently successful, sinking 10 out of 10 shots with ease. Others, however, struggled to make even three or four baskets out of 10. This disparity wasn't due to physical limitations—every player had the strength and coordination needed to succeed.

The difference lay in their ability to regulate their emotions. I watched one player who struggled to focus, distracted by the noise in the stadium and his own internal chatter. His Emotion Regulation Circuit wasn't able to manage the turbulence in his brain.

Greg's training had a profound impact on the Beavers' performance, leading to unprecedented success. Under his guidance, the baseball team won their league twice in a row and clinched the College World Series in 2018.

When athletes learn to turn on the Emotion Regulation Circuit, they prevent anxiety and stress from undermining their abilities. Just as seasoned meditators quiet the emotional limbic system with the ventromedial prefrontal cortex, then boost focus by activating the orbitofrontal cortex, athletes can do the same. At the highest levels of competition, emotion regulation and the power to focus can be more crucial than physical prowess.

7.19. The Oregon State Beavers in action.

How Much Spiritual Intelligence Is Required for Change?

The previous projections show that we're on the verge of a jump to a much more compassionate, peaceful, and creative society, one in which human potential is unleashed at every level. We also see the countertrends in consciousness, where 20% of people are increasingly gloomy about their circumstances despite the overwhelming objective evidence that we're already centuries into an epoch of unprecedented human thriving.

How many people will need to turn on the four circuits of the awakened brain and run the software of SQ in order for massive social change to occur? The answer is: fewer than you might think.

Harvard's Erica Chenoweth and colleagues have conducted extensive research on peaceful change.[75] Their studies reveal that nonviolent social movements are twice as likely to succeed as violent ones. A key finding from their research is the 3.5% rule: *When at least 3.5% of a population engages in nonviolent protest, the movement has never failed to bring about significant change.*[76] This statistic underscores the power of collective, peaceful action in driving social and political transformation.

7.20. Manila's prayer and protest gathering.

In 1972, to shore up his rule, Philippine President Ferdinand Marcos declared martial law. This led to widespread human rights abuses, suppression of dissent, and corruption. In 1986, more than two million Filipinos gathered for prayer and peaceful protest in the streets of Manila. This represented a significant portion of the population, well over the 3.5% threshold. The Marcos regime folded on the fourth day.

In the 1960s, the civil rights movement, inspired by Martin Luther King Jr., engaged in sit-ins and marches to challenge racial segregation. Engaging over 3.5% of the population of the US Southern states, it changed the national mind, leading to the Civil Rights Act of 1964 and the Voting Rights Act of 1965.

A large-scale review examined the effect of changes in consciousness on social wellbeing.[77] The researchers focused on a five-year period in which the number of people using Transcendental Meditation (TM) in intentional groups exceeded *the square root of 1% of the US population.* That's the threshold at which they found effects—a number far smaller than the 3.5% identified in the Harvard studies.

They found substantial reductions in several indicators of national stress. These included a 24% reduction in violent crime, a 13.8% decrease in suicides, an 11.5% decrease in automobile accident deaths, and a 15.7% reduction in the unemployment rate. In the Extended Play section, you'll find an inspiring audio interview with Dr. Tony Nader, the worldwide head of TM.

Whether it's 3.5% or the square root of 1% changing their consciousness, it's clear that *it doesn't take more than a small percentage of a population to create massive social transformation.* The changes of our collective mind produced by SQ have been creating wellbeing in society for generations, accelerating with the growing number of meditators. SQ has effects that reach far beyond the individual level to create societal peace and stability. By cultivating SQ, you're contributing to the critical mass of people needed to elevate the consciousness of the planet.

Illuminating the Future

As we enter the next era in human evolution, we recognize that *SQ produces not just individual wellbeing but collective awakening.* The journey of ideas in this book has shown us that enlightenment is not a distant dream reserved for mystics but a tangible reality embedded in the very fabric of our brains. Every one of us is wired for ecstasy, for connection, and for resonance with the universe.

Our brains contain structures essential for survival, yet they are also built to transcend the mundane and participate in the divine. As well as beings of impulse and conditioning, we are equally beings of light, capable of illuminating not just our own lives but the world around us.

The rapid advances in neuroscience have revealed that the cultivation of SQ is not just personally enriching but a catalyst for global transformation. As we nurture the circuits of the Enlightenment Network, we are not just changing our individual experience; we are contributing to a larger evolutionary leap.

Each meditation, each act of compassion, each moment of flow, each word of kindness, ripples out into the collective consciousness, accelerating the pace of human flourishing. The time has come for us to recognize that our personal growth is inextricably linked to the wellbeing of our planet and the future of our species.

Imagine a world where SQ is as commonplace as literacy, where children are taught from an early age to access their inner wisdom and connect with the one great universal consciousness. This is not a utopian fantasy, but a realistic vision grounded in the scientific breakthroughs we have explored together.

As more people awaken to their spiritual potential, we will see a profound shift in every area of human endeavor. From politics to education, from healthcare to the arts, the ripple effects of a more enlightened population will create a more compassionate, creative, joyful, and resilient world.

This vision is not just a possibility; it is an imperative. The challenges we face as a planet—climate change, social inequality, pollution, political turmoil, repression, intolerance, species extinction, health threats—menace everything the human race has created over the past millennia.

Yet SQ treats them not just as problems to solve but *as opportunities to evolve*. SQ offers us the tools to meet these challenges with wisdom and grace, to transform adversity into growth, and to co-create a future where all beings can thrive.

The science is clear: the more we cultivate our SQ, the more we will be able to navigate the complexities of the modern world with clarity, purpose, wisdom, and love.

Please join me in embracing SQ not just as a personal quest but as our collective destiny. The world needs more Finders, more people who have crossed the tipping point and live from a place of fundamental wellbeing. Let us commit to being the pioneers of this new frontier, the torchbearers of a bright and enlightened future.

As we do, we will discover that the true power of SQ lies not just in the extraordinary, but in the everyday moments of connection, compassion, and joy we share. Together, we will illuminate the path forward for ourselves, for each other, for the planet, and for generations to come.

Deepening Practices

The Deepening Practices for this chapter include:

- **Journaling Exercise 1:** How will living in a world of spiritually intelligent people affect your life? In the domains of work, money, relationships, and health?
- **Journaling Exercise 2:** How long a lifespan do you anticipate for yourself? Write down the first number that comes into your head. Now describe the details of your most inspiring vision for the last 10 years of your life.
- **Visualization:** When you meditate this week, imagine the millions of people around the world that are meditating at that very same time. Check the Insight Timer app and find out the number meditating that day.
- Dawson Guided Neuroliminal: Flourishing World, Flourishing Brain

Get the meditation track at: SpiritualIntelligence.info/7.

Extended Play Resources

The Extended Play version of this chapter includes:

- Peace events of the 20th and 21st centuries listed by decade
- The Trillion Tree Project
- Podcast: Reversing Alzheimer's: Dr. Heather Sandison and Dawson Church in conversation
- Searchable database of energy therapy studies
- Podcast: Consciousness Is All There Is: Dr. Tony Nader, worldwide head of TM, and Dawson Church in conversation

Get the extended play resources at: SpiritualIntelligence.info/7.

Updates to This Chapter

As new studies are published, this chapter is regularly updated. Get the most recent version at: SpiritualIntelligence.info/7.

About the Writing of Spiritual Intelligence

The Voyage of Discovery

Spiritual Intelligence is the fifth in a series of books I have written about the intersection of mind, brain, spirituality, intention, energy, healing, and the future of human evolution.

The first book was called *Soul Medicine,* published in 2004. It was coauthored with my mentor, Norm Shealy, MD, PhD, after completing my dissertation at Holos University. Norm and a group of visionary academics founded Holos in 2001 and brought academic rigor to the study of energy medicine and alternative therapies.

Soul Medicine is about the application of energy to healing and it's full of startling revelations. Norm was a medical genius, a visionary, and a walking encyclopedia of knowledge about the subject. I was privileged to have Norm as a warm and generous mentor for the last two decades of his life.

The second book was *The Genie in Your Genes,* published in 2007. I was excited about the potential of energy therapies to change gene expression. I noticed that professionals in the fields of energy medicine and energy psychology knew nothing about epigenetics, and most biologists knew little about the role that energy played in healing. I pulled together all the relevant information from disparate fields of science and clinical practice to show that intangible influences like belief, thought, intention, and energy were epigenetic, changing the expression of genes, hormones, neurotransmitters, enzymes, and various molecules inside our bodies. The book has gone through several editions and I update it with new research regularly, though the underlying assumptions have stood the test of decades.

The third book was *Mind to Matter,* published in 2018. I was intrigued by the research showing the connection between thoughts and things. I set out to find the scientific evidence for every link in the chain between us having a thought and it manifesting as "objective reality" in the world around us. I was astonished at how much scientific evidence there was for the link and surprised that no one had pieced it all together before.

That led to *Bliss Brain,* the fourth book, published in 2020. While writing about Tibetan monks, Franciscan nuns, and other spiritual adepts in my previous books, I had begun to use their practices. I found myself getting happier and happier, with my hour of meditation being the most precious time of my day.

In 2017, I experienced a series of tragedies in my life. The first was the sudden incineration of my house and office in a massive fast-moving wildfire. The following year brought physical illness, the crash of my business, and the loss of my life's savings.

But as a long-time meditator, I found myself happy regardless of external circumstances. As I looked into the characteristics of neurotransmitters such as anandamide, the bliss molecule, and the brain regions active in mystics, I discovered the physiological underpinnings of unconditional ecstasy and shared them in *Bliss Brain.*

The most remarkable finding is the way the brain changes under the influence of ecstatic states. I began to wonder about the long-term implications of this for human evolution. I realized that Spiritual Intelligence, a term coined in the 1990s, provided a framework for tying all these threads together.

My early outline of *Spiritual Intelligence* in 2019 was based on my assumption that this intelligence (abbreviated SQ, as in spiritual IQ) was a spiritual and mental phenomenon. Excellent definitions of SQ come from the academics who have studied it. They generally describe it as the cognitive capacity to find meaning and purpose in life and establish a transcendent relationship with something larger than oneself. That was my starting point too.

But as I dug deeper into the science, I realized that I was completely wrong. SQ is based squarely on physiology: the growth of specific regions of the brain. Reading the writings of the mystics, I also realized that SQ is not just found in the mind. It's body- and emotion-based too.

In Patanjali's *Yoga Sutras,* he has students of enlightenment start by learning the physical poses of Hatha Yoga. They then learn *pranayama* or breathwork. Only then does he consider them ready for meditation. And only after extensive meditation on simple objects like a candle, yantra. or mantra do they approach the elevated states of Samadhi. Even then, Patanajali teaches aspirants a set of *physical* exercises to transport their awareness to higher states.

As I realized how body-based SQ was, how dependent it was on the growth of neurons in the four circuits of the Enlightenment Network, I had to ditch my whole premise for *Spiritual Intelligence* and start again. Because of a demanding work schedule that gives me only small windows for writing, it took till 2025 to complete the text and the extensive resources in the Extended Play sections.

While genres like nonfiction and memoir can be written quickly, writing a science book is a different proposition. Sometimes I need to read 10 or 20 studies to find the one that is most relevant. It can take two or three days of evaluating, understanding, and finding lay analogies to academic concepts to write just half a page.

Every fact in a popular science book like the ones discussed here has to be scrupulously accurate. While I can write a nonfiction essay, a personal reflection, or an inspirational blog post in a few hours, a chapter for a book like these requires hundreds of hours to compose and refine.

There was a great deal more material I wanted to include in *Spiritual Intelligence,* but it would have made the book too long. So I cut about half the chapters I'd originally planned to include. I also cut material from every chapter. SQ is already a big topic for readers to wrap their minds around and I wanted to make the book approachable. For a writer, cutting is painful, but the experience of the reader is the top priority.

My Intentions in Writing This Book

I had several intentions in writing *Spiritual Intelligence*. My primary goal was to contribute to the growth of cultural awareness of SQ.

The term "spiritual intelligence" was coined in 1997 by business consultant Dana Zohar. Several books on the subject were published in the late 1990s and early 2000s. Cindy Wigglesworth developed the Spiritual Intelligence Inventory in 2000. The first handful of studies of SQ appeared in the early 21st century, with more emerging in subsequent decades, leading to a growing body of academic research on the subject.

However, the concept of SQ failed to permeate the culture and shift the social conversation in the way Emotional Intelligence or EQ did in the mid-1990s.

Today, conditions are different due to the growth in meditation and the movement of millions of people away from organized religion and into the "spiritual but not religious" category. My primary aim in this book is to kindle popular awareness of SQ in mass culture.

My second goal is to make it clear that everyone has the neurological hardwiring for SQ. It isn't an esoteric or mystical state, but a fundamental characteristic of the human brain.

This scientific fact, popularized in the books cited in the first three chapters, is a game-changer. No one needs to go somewhere to "find" SQ—it's inherent in the structure of our brains. This shifts the pursuit of SQ from groping through the maze of competing religions and spiritual traditions into an empirical exercise in neuron-building. My expectation is that we will approach building spiritual intelligence—and the brain health that follows—in the same methodical way we now strengthen our muscles through exercise.

My third goal in writing this book is to provide a glimpse into the interior world of the mystic, sharing the ecstasy that comes from transcendent experiences. Everyone wants to be happy, and spiritual experience is a quick and certain path to that happiness. By showing just how joyful we can become, I hope to inspire millions of people to focus on developing their SQ.

My fourth goal is to highlight the science that drives this quest. While our ancestors had the priceless touchstones of written scripture and the stories of mystics, we now have MRIs, EEGs, gene assays, apps, AI, VR, and other advanced technologies. These tools show us which practices develop SQ most efficiently, accelerating spiritual development for millions of people and for the planet as a whole.

Finally, my goal is to extrapolate from today's knowledge to show where SQ is taking us in the future. The road ahead is not just hopeful—it's transformative. We are in the midst of an unprecedented evolutionary leap and its implications will touch every field of human endeavor. The future of the planet is being irretrievably altered by the influence of SQ on the human brain. My hope is that this vision will dispel the unwarranted gloom in public discourse and help us find a fact-based, practical path to a better world.

Acknowledgments

I am beyond grateful to the many people who have played a part in the creation of this book. First, to the many pioneers who have promoted our evolving understanding of SQ over the past three decades.

Second, to my wonderful team at EFT Universe. Without them running the business, I would not have had the time off work to write the book.

Third, to people who have contributed concepts and ideas through their books and research. I read and reread—with astonishment—page 78 of *Altered Traits* by Richard Davidson and Daniel Goleman. On that page they describe the four distinct circuits of the Enlightenment Network. Others like Andrew Newberg in *How Enlightenment Changes Your Brain* and Lisa Miller in *The Awakened Brain* develop the idea of brain regions involved in awakening, but, as far as I am aware, it was *Altered Traits* that first identified these four specific circuits.

I appreciate people who read sections of the manuscript and provided comments or with whom I talked in my effort to educate myself about SQ. These included Andrew Vidich, Jeff Cox, Cindy Wigglesworth, Jean Houston, Donna Bach, Gary Groesbeck, Judith Pennington, Jeffery Martin, Fred Travis, David Gandelman, Ervin Laszlo, Bob Hoss, Yosi Amram, Steve Taylor, Marci Shimoff, Sergio Barone, Kristin Kirk, Sadhvi Saraswati, and Stephen Altair.

Finally, I appreciate those at Energy Psychology Press who have worked on taking *Spiritual Intelligence* from a manuscript to a book. These include Stephanie Marohn, who copyedited the text, Karin Kinsey, who typeset the book and created the illustrations, Vicki Valentine, who designed the interior pages and cover, Anitha Vasudevan, who managed the print bidding and book manufacturing process, and our distributor Baker & Taylor.

I am also indebted to the many distinguished colleagues and best-selling authors who were kind enough to review early versions of the manuscript and provide endorsements. When a very busy person spends time with more than 300 pages, then spends further time crafting a testimonial, they are giving a very generous gift indeed.

My Use of Artificial Intelligence

Marketer Seth Godin compares AI to a mediocre writer and editor. Its output is bland. But it can collect many resources and do a basic job quickly.

Before AI became widely available in 2022, I tasked human editors with producing initial drafts. I would ask my editor Stephanie Marohn or my project manager Anitha Vasudevan to write a section on a study or a story of a celebrity. I would then develop that draft. An example of this is the story of Dwayne Johnson at the start of Chapter 2 or the Alzheimer's study that concludes the chapter.

After I began using ChatGPT-4 in 2023, I made similar use of AI in the second half of the book. AI sped up the creation of content by creating basic factual text quickly and efficiently. Rather than drawing on many hours of Stephanie or Anitha's time that could profitably be devoted to other high-leverage projects, I could have a draft in under a minute. I would then rewrite the material to address AI's deficiencies.

In the Extended Play resources on the book's web site, you can find an example of an AI draft, followed by a discussion of the many limitations and inaccuracies inherent in AI, such as dull and pedantic prose and fictitious references, and how I worked around them. To indicate where AI was used in image generation, in the Image Rights lists I attribute an image to "Author AI" rather than simply "Author."

Updates to This Text

Every science book is obsolete by the time it is published. To allow time for copyediting, design, typesetting, printing, and distribution, there's about a year's gap between completion of the manuscript and the appearance of the printed copy. During that year, new studies are published in peer-reviewed journals, updating the knowledge base for the book's subject.

The traditional way to address this is for the author to write a new edition every few years. That's what I've done for my previous books. However, the web now allows a fresh approach and I've used it here. I can drop updates onto the book's website and keep the text current with the latest research.

The basic premises of the original book have to be sound in order for this to work. I'm gratified that even decades after *The Genie in Your Genes* was first published, its fundamental arguments have stood the test of time. My hope is that this will be as true of *Spiritual Intelligence* as it has been for my earlier books.

Future Directions

The first two hours of each day, spent in meditation, reading, and journaling, are the cornerstone of my life. I also take a few days off work each month, as well as the first couple of months of each year. During these times, I enter Samadhi. After that time off ends, I ground myself and joyfully reenter the world of "chop wood, carry water."

I plan to keep on offering in-person and virtual classes and retreats. I also enjoy sharing the joy and potential of awakening with others, so I'll continue appearing on podcasts and summits for as long as I'm able. I love deep conversations with other Awakeners and Finders, and I look forward to immersing myself in these soul connections for the rest of my life.

Seeing the remarkable transformations of participants in the Short Path to Oneness program, A 21 Day Walk With Your Higher Power, EFT workshops, and our other courses, is a constant source of wonder for me. I am so grateful to science for allowing modern humans to identify the most efficient tools from ancient traditions and apply them for rapid and reliable change. Several new books on these topics are percolating.

I also enjoy connecting with my many family members, including my and Christine's children and grandchildren, and look forward to continuing to play a part in their lives till I drop.

My newsletters at EFT Universe and UpLife are the ways in which I stay in touch with people, so please subscribe at DawsonGift.com if you aren't already part of our community.

References

Chapter 1

1. Equal Justice Initiative. (2022). *Anthony Ray Hinton: Mr. Hinton spent 30 years on death row for a crime he did not commit.* Retrieved from https://eji.org/cases/anthony-ray-hinton
2. Church, D. (2015). *Psychological trauma: Healing its roots in brain, body, and memory.* Energy Psychology Press.
3. Pickrell, J. (2019). How the earliest mammals thrived alongside dinosaurs. *Nature, 574*(7779), 468-473.
4. Handwork, B. (2021). An evolutionary timeline of Homo Sapiens. *Smithsonian Magazine.* Retrieved from https://www.smithsonianmag.com/science-nature/essential-timeline-understanding-evolution-homo-sapiens-180976807
5. Du, A., Zipkin, A. M., Hatala, K. G., Renner, E., Baker, J. L., Bianchi, S., ...Wood, B. A. (2018). Pattern and process in hominin brain size evolution are scale-dependent. *Proceedings of the Royal Society B: Biological Sciences, 285*(1873), 20172738.
6. Church, D. (2020). *Bliss brain: The neuroscience of remodeling your brain for resilience, creativity, and joy.* Hay House.
7. *Ibid.*
8. Guy-Evans, O. (2021). What does the brain's cerebral cortex do? *SimplyPsychology.* Retrieved from https://www.simplypsychology.org/what-is-the-cerebral-cortex.html
9. Herculano-Houzel, S. (2009). The human brain in numbers: A linearly scaled-up primate brain. *Frontiers in Human Neuroscience, 3,* 31.
10. Glickstein, M., & Doron, K. (2008). Cerebellum: Connections and functions. *Cerebellum, 7*(4), 589-594.
11. Timmermann, A. (2020). Quantifying the potential causes of Neanderthal extinction: Abrupt climate change versus competition and interbreeding. *Quaternary Science Reviews, 238,* 106331.
12. McCauley B. (2019). Life expectancy in hunter-gatherers. In T. Shackelford & V. Weekes-Shackelford (Eds.), *Encyclopedia of evolutionary psychological science.* Springer.
13. Volk, A. A., & Atkinson, J. A. (2013). Infant and child death in the human environment of evolutionary adaptation. *Evolution and Human Behavior, 34*(3), 182-192.
14. Roser, M. (2019). Mortality in the past—around half died as children. *Our World in Data.* Retrieved from https://ourworldindata.org/child-mortality-in-the-past
15. Cold Spring Harbor Laboratory. (2020). Chromosome 10: Gene which creates cortisol, Matt Ridley (interview). *DNA Learning Center.* Retrieved from https://dnalc.cshl.edu/view/15405-Chromosome-10-gene-which-creates-cortisol-Matt-Ridley.html
16. National Library of Medicine. (2021). CYP17A1 gene. *MedlinePlus.* Retrieved from https://medlineplus.gov/genetics/gene/cyp17a1/.
17. Pinker, S. (2018). *Enlightenment now: The case for reason, science, humanism, and progress.* Penguin UK.
18. Reddan, M. C., Wager, T. D., & Schiller, D. (2018). Attenuating neural threat expression with imagination. *Neuron, 100*(4), 994-1005.
19. Killingsworth, M. A., & Gilbert, D. T. (2010). A wandering mind is an unhappy mind. *Science, 330*(6006), 932-932.
20. Liang, X., Zou, Q., He, Y., & Yang, Y. (2013). Coupling of functional connectivity and regional cerebral blood flow reveals a physiological basis for network hubs of the human brain. *Proceedings of the National Academy of Sciences, 110*(5), 1929-1934.
21. Boyatzis, R. E., Rochford, K., & Jack, A. I. (2014). Antagonistic neural networks underlying differentiated leadership roles. *Frontiers in Human Neuroscience, 8,* 114.
22. Shi, L., Sun, J., Wu, X., Wei, D., Chen, Q., Yang, W., . . . Qiu, J. (2018). Brain networks of happiness: Dynamic functional connectivity among the default, cognitive and salience networks relates to subjective well-being. *Social Cognitive and Affective Neuroscience, 13*(8), 851-862.
23. Vaish, A., Grossmann, T., & Woodward, A. (2008). Not all emotions are created equal: The negativity bias in social-emotional development. *Psychological Bulletin, 134*(3), 383-403.

24. Gillespie, C., Szabo, S., & Nemeroff, C. B. (2020). Unipolar depression. In R. N., Rosenberg & J. M. Pascual (Eds.), *Rosenberg's molecular and genetic basis of neurological and psychiatric disease* (6th Ed.). Elsevier.
25. Raichle, M. E. (2015). The brain's default mode network. *Annual Review of Neuroscience, 38,* 433-447.
26. Killingsworth, M. A., & Gilbert, D. T. (2010). A wandering mind is an unhappy mind. *Science, 330*(6006), 932.
27. Cummins, R. A. (2012). Positive psychology and subjective well-being homeostasis: A critical examination of congruence. In A. Efklides & D. Moraitou (Eds.), *A positive psychology perspective on quality of life* (pp. 67-86). Springer, Dordrecht.
28. Hasenkamp, W., Wilson-Mendenhall, C. D., Duncan, E., & Barsalou, L. W. (2012). Mind wandering and attention during focused meditation: A fine-grained temporal analysis of fluctuating cognitive states. *Neuroimage, 59*(1), 750-760.
29. Santomauro, D. F., Herrera, A. M. M., Shadid, J., Zheng, P., Ashbaugh, C., Pigott, D. M.,... Ferrari, A. J. (2021). Global prevalence and burden of depressive and anxiety disorders in 204 countries and territories in 2020 due to the COVID-19 pandemic. *Lancet, 398*(10312), 1700-1712.
30. Huxley, A. (2009). *The perennial philosophy.* Harper Perennial.
31. Lipton, B. H. (2005). *The biology of belief: Unleashing the power of consciousness, matter, and miracles.* Hay House.
32. Church, D. (2014). *The genie in your genes: Epigenetic medicine and the new biology of intention* (3rd ed.). Energy Psychology Press.
33. Church, D., Yount, G., Rachlin, K., Fox, L., & Nelms, J. (2018). Epigenetic effects of PTSD remediation in veterans using clinical emotional freedom techniques: A randomized controlled pilot study. *American Journal of Health Promotion, 32*(1), 112-122.
34. Norris, C. J. (2021). The negativity bias, revisited: Evidence from neuroscience measures and an individual differences approach. *Social Neuroscience, 16*(1), 68-82.
35. Church, D. (2018). Mind to matter: *The astonishing science of how your brain creates material reality.* Hay House.
36. Church, D. (2020). *Bliss brain: The neuroscience of remodeling your brain for resilience, creativity, and joy.* Hay House.
37. Church, D., Stapleton, P., & Sabot, D. (2020). Brief EcoMeditation associated with psychological improvements: A preliminary study. *Global Advances in Health and Medicine, 9,* 2164956120984142.
38. Pennington, J., Sabot, D., & Church, D. (2019). EcoMeditation and Emotional Freedom Techniques (EFT) produce elevated brain-wave patterns and states of consciousness. *Energy Psychology: Theory, Research, and Treatment, 11*(1), 13-40.
39. Church, D., Stapleton, P., Baumann, O., & Sabot, D. (2022). *EcoMeditation modifies brain resting state network activity. Innovations in Clinical Neuroscience, 19*(7-9), 61-70.
40. Goleman, D. (1995). *Emotional intelligence: Why it can matter more than IQ.* Bantam.
41. Goleman, D., & Davidson, R. J. (2018). *Altered traits: Science reveals how meditation changes your mind, brain, and body.* Penguin.
42. Newberg, A., & Waldman, M. R. (2017). *How enlightenment changes your brain: The new science of transformation.* Penguin.
43. *Ibid.*
44. Nuwer, R. (2012). The world's happiest man is a Tibetan monk. *Smithsonian Magazine.* Retrieved from https://www.smithsonianmag.com/smart-news/the-worlds-happiest-man-is-a-tibetan-monk-105980614
45. Church, D., Stapleton, P., Baumann, O., & Sabot, D. (2022). EcoMeditation modifies brain resting state network activity. *Innovations in Clinical Neuroscience, 19*(7-9), 61-70.
46. Pennington, J., Sabot, D., & Church, D. (2019). EcoMeditation and Emotional Freedom Techniques (EFT) produce elevated brain-wave patterns and states of consciousness. *Energy Psychology: Theory, Research, and Treatment, 11*(1), 13-40.
47. Yogananda, P. (2019). *Autobiography of a yogi.* MJF Books.
48. Singh, K. (1968). *A great saint: Baba Jaimal Singh, his life and teachings* (2nd ed.). Ruhani Satsang.

49. Fritz & Fehmi. (1982). *The Open Focus handbook: The self regulation of attention in biofeedback training and everyday activities.* Biofeedback Computers.
50. Leyton, M., Stewart, S., & Canadian Centre on Substance Abuse. (2014). *Substance abuse in Canada: Childhood and adolescent pathways to substance use disorders* (Fig. 7). Retrieved from https://www.researchgate.net/figure/Brain-circuits-involved-in-the-regulation-of-emotions-and-processing-of-rewards_fig4_283732455
51. Khalaf, O., Resch, S., Dixsaut, L., Gorden, V., Glauser, L., & Gräff, J. (2018). Reactivation of recall-induced neurons contributes to remote fear memory attenuation. *Science, 360*(6394), 1239-1242.
52. Viviani, R. (2014). Neural correlates of emotion regulation in the ventral prefrontal cortex and the encoding of subjective value and economic utility. *Frontiers in Psychiatry, 5,* 123.
53. Bush, G. (2010). Attention-deficit/hyperactivity disorder and attention networks. *Neuropsychopharmacology, 35*(1), 278-300.
54. Kim, S., Reeve, J. M., & Bong, M. (2016). Introduction to motivational neuroscience. In S. Kim, J. M. Reeve & M. Bong (Eds.), *Recent developments in neuroscience research on human motivation* (Vol. 19 of *Advances in Motivation and Achievement* book series; pp. 1-19, Fig. 5). Emerald Group.
55. Mason, M. F., Norton, M. I., Van Horn, J. D., Wegner, D. M., Grafton, S. T., & Macrae, C. N. (2007). Wandering minds: The default network and stimulus-independent thought. *Science, 315*(5810), 393-395.
56. Martin, J. A. (2019). *The finders.* Integration Press.
57. Decety, J., Bartal, I. B. A., Uzefovsky, F., & Knafo-Noam, A. (2016). Empathy as a driver of prosocial behaviour: Highly conserved neurobehavioural mechanisms across species. *Philosophical Transactions of the Royal Society B: Biological Sciences, 371*(1686), 20150077.
58. Namkung, H., Kim, S. H., & Sawa, A. (2017). The insula: An underestimated brain area in clinical neuroscience, psychiatry, and neurology. *Trends in Neurosciences, 40*(4), 200-207.
59. Blakeslee, S. (2007). A small part of the brain, and its profound effects. *New York Times.* Retrieved from https://www.nytimes.com/2007/02/06/health/psychology/06brain.html
60. Lutz, A., Brefczynski-Lewis, J., Johnstone, T., & Davidson, R. J. (2008). Regulation of the neural circuitry of emotion by compassion meditation: Effects of meditative expertise. *PloS One, 3*(3), e1897.
61. Goleman, D., & Davidson, R. J. (2018). *Altered traits: Science reveals how meditation changes your mind, brain, and body.* Penguin.
62. Cheng, X., Yuan, Y., Wang, Y., & Wang, R. (2020). Neural antagonistic mechanism between default-mode and task-positive networks. *Neurocomputing, 417,* 74-85.
63. Vaish, A., Grossmann, T., & Woodward, A. (2008). Not all emotions are created equal: The negativity bias in social-emotional development. *Psychological Bulletin, 134*(3), 383-403.
64. Cherry, K. (2021). Gardner's theory of multiple intelligences. *Verywellmind.* Retrieved from https://www.verywellmind.com/gardners-theory-of-multiple-intelligences-2795161
65. Chotiner, I. (2018). How to survive death row. *Slate.* Retrieved from https://slate.com/news-and-politics/2018/03/how-anthony-ray-hinton-survived-death-row.html
66. Koch, C. (2018). What is consciousness? *Nature, 557,* S8-S12.
67. Martschenko, D. (2018). The IQ test wars: Why screening for intelligence is still so controversial. *The Conversation.* Retrieved from https://theconversation.com/the-iq-test-wars-why-screening-for-intelligence-is-still-so-controversial-81428
68. King, D. B., & DeCicco, T. L. (2009). A viable model and self-report measure of spiritual intelligence. *The International Journal of Transpersonal Studies, 28*(1), 68-85.
69. Bit na, D. K., & M rtinsone, K. (2021). Mystical experience has a stronger relationship with spiritual intelligence than with schizotypal personality traits and psychotic symptoms. *Psychology of Consciousness: Theory, Research, and Practice, 11*(2), 137-153.

Chapter 2

1. Gallo, C. (2022). Three reasons Dwayne Johnson's 'Seven Bucks' story is worth repeating. *Forbes.* Retrieved from https://www.forbes.com/sites/carminegallo/2022/10/20/three-reasons-dwayne-johnsons-seven-bucks-story-is-worth-repeating/?sh=7771ede150d6

2. Barcelona, A. (2021). Kind List 2021: The Rock. *Hello!* Retrieved from https://www.hellomagazine.com/celebrities/20211112124494/dwayne-johnson-the-rock-kind-list-2021
3. Showering Love. (2020). Dwayne Johnson—A "Rock" to the community. *Showering Love.* Retrieved from https://showeringlove.org/dwayne-johnson-a-rock-to-the-community
4. Institute of Medicine and National Academy of Sciences. (1992). *Discovering the brain.* National Academies Press. Retrieved from https://www.nap.edu/read/1785/chapter/7
5. Abdissa, D., Hamba, N., & Gerbi, A. (2020). Review article on adult neurogenesis in humans. *Translational Research in Anatomy, 20,* 100074.
6. Cole, J. D., Espinueva, D. F., Seib, D. R., Ash, A. M., Cooke, M. B., Cahill, S. P.,...Snyder, J. S. (2020). Adult-born hippocampal neurons undergo extended development and are morphologically distinct from neonatally-born neurons. *Journal of Neuroscience, 40*(30), 5740-5756.
7. Garcia, K. E., Kroenke, C. D., & Bayly, P. V. (2018). Mechanics of cortical folding: stress, growth and stability. *Philosophical Transactions of the Royal Society B: Biological Sciences, 373*(1759), 20170321. Retrieved from https://royalsocietypublishing.org/doi/10.1098/rstb.2017.0321
8. *Ibid.*
9. Lucassen, P. J., Fitzsimons, C. P., Salta, E., & Maletic-Savatic, M. (2020). Adult neurogenesis, human after all (again): Classic, optimized, and future approaches. *Behavioural Brain Research, 381,* 112458.
10. Gregory, M. D., Kippenhan, J. S., Dickinson, D., Carrasco, J., Mattay, V. S., Weinberger, D. R., & Berman, K. F. (2016). Regional variations in brain gyrification are associated with general cognitive ability in humans. *Current Biology, 26*(10), 1301-1305.
11. Harvard Medical School. (2016). *Preserve your muscle mass.* Retrieved from https://www.health.harvard.edu/staying-healthy/preserve-your-muscle-mass
12. WebMD. (2021). *Which area of the brain is most susceptible to shrinkage as we age?* Retrieved from https://www.webmd.com/healthy-aging/which-area-of-the-brain-is-most-suscepitble-to-shrinkage-as-we-age
13. Peters, R. (2006). Ageing and the brain. *Postgraduate Medical Journal, 82*(964), 84-88.
14. WebMD. (2021). *Which area of the brain is most susceptible to shrinkage as we age?* Retrieved from https://www.webmd.com/healthy-aging/which-area-of-the-brain-is-most-suscepitble-to-shrinkage-as-we-age
15. Dementia Care Central. (n.d). *Normal brain vs. Alzheimer's.* Retrieved from https://www.dementiacarecentral.com/video/video-brain-changes
16. *Ibid.*
17. Bayley, J. (1998). Elegy for Iris: Scenes from an indomitable marriage. *New Yorker.* Retrieved from https://www.newyorker.com/magazine/1998/07/27/elegy-for-iris
18. *Ibid.*
19. Nicholls, R. (1999). Iris Murdoch, novelist and philosopher, is dead. *New York Times.* Retrieved from https://www.nytimes.com/1999/02/09/books/iris-murdoch-novelist-and-philosopher-is-dead.html
20. Bayley, J. (1999). *Elegy for Iris.* St. Martin's Press.
21. Moreno-Jiménez, E. P., Flor-García, M., Terreros-Roncal, J., Rábano, A., Cafini, F., Pallas-Bazarra, N.,...Llorens-Martín, M. (2019). Adult hippocampal neurogenesis is abundant in neurologically healthy subjects and drops sharply in patients with Alzheimer's disease. *Nature Medicine, 25*(4), 554-560.
22. Yoon, Y. B., Shin, W. G., Lee, T. Y., Hur, J. W., Cho, K. I. K., Sohn, W. S.,...Kwon, J. S. (2017). Brain structural networks associated with intelligence and visuomotor ability. *Scientific Reports, 7*(1), 1-9.
23. Goriounova, N. A., & Mansvelder, H. D. (2019). Genes, cells and brain areas of intelligence. *Frontiers in Human Neuroscience, 13,* 44.
24. O'Neil, L. (2013). The brain and emotional intelligence by Daniel Goleman. *Learning and the Brain.* Retrieved from https://www.learningandthebrain.com/blog/the-brain-and-emotional-intelligence-by-daniel-goleman
25. Singer, T. (2022). *Research focus.* Retrieved from https://www.social.mpg.de/64505/publications

26. Sahay, A., & Hen, R. (2007). Adult hippocampal neurogenesis in depression. *Nature Neuroscience, 10*(9), 1110-1115.
27. Bir, S. C., Ambekar, S., Kukreja, S., & Nanda, A. (2015). Julius Caesar Arantius (Giulio Cesare Aranzi, 1530–1589) and the hippocampus of the human brain: History behind the discovery. *Journal of Neurosurgery, 122*(4), 971-975.
28. Dunea, G. (2020) Julius Caesar Aranzi, anatomist and surgeon of Bologna. *Hektoen International Journal, Summer*. Retrieved from https://hekint.org/2020/07/20/julius-caesar-aranzi-anatomist-and-surgeon-of-bologna
29. Rice University. (n.d.). *Hippocampus*. Retrieved from https://www.caam.rice.edu/~cox/wrap/hippocampus.pdf
30. Mannekote Thippaiah, S., Pradhan, B., Voyiaziakis, E., Shetty, R., Iyengar, S., Olson, C., & Tang, Y. Y. (2022). Possible role of parvalbumin interneurons in meditation and psychiatric illness. *Journal of Neuropsychiatry and Clinical Neurosciences, 34*(2), 113-123.
31. Noguchi, H., Murao, N., Kimura, A., Matsuda, T., Namihira, M., & Nakashima, K. (2016). DNA methyltransferase 1 is indispensable for development of the hippocampal dentate gyrus. *Journal of Neuroscience, 36*(22), 6050-6068.
32. Bremner, J. D. (2006). Traumatic stress: effects on the brain. *Dialogues in Clinical Neuroscience, 8*(4), 445-461.
33. Luders, E., Cherbuin, N., & Kurth, F. (2015). Forever Young (er): Potential age-defying effects of long-term meditation on gray matter atrophy. *Frontiers in Psychology, 5,* 1551.
34. Luders, E., Kurth, F., Mayer, E. A., Toga, A. W., Narr, K. L., & Gaser, C. (2012). The unique brain anatomy of meditation practitioners: Alterations in cortical gyrification. *Frontiers in Human Neuroscience, 6,* 34.
35. Markel, H. (2023). The strange story of Einstein's brain. *PBS News Hour*. Retrieved from https://www.pbs.org/newshour/health/the-strange-story-of-einsteins-brain
36. Men, W., Falk, D., Sun, T., Chen, W., Li, J., Yin, D.,...Fan, M. (2014). The corpus callosum of Albert Einstein's brain: Another clue to his high intelligence? *Brain, 137*(4), e268-e268.
37. Chan, R. W., Leong, A. T., Ho, L. C., Gao, P. P., Wong, E. C., Dong, C. M., . . . Wu, E. X. (2017). Low-frequency hippocampal–cortical activity drives brain-wide resting-state functional MRI connectivity. *Proceedings of the National Academy of Sciences, 114*(33), E6972-E6981.
38. Takimoto-Ohnishi, E., Ohnishi, J., & Murakami, K. (2012). Mind-body medicine: Effect of the mind on gene expression. *Personalized Medicine Universe, 1*(1), 2-6.
39. Innes, K. E., Selfe, T. K., Khalsa, D. S., & Kandati, S. (2016). Effects of meditation versus music listening on perceived stress, mood, sleep, and quality of life in adults with early memory loss: A pilot randomized controlled trial. *Journal of Alzheimer's Disease, 52*(4), 1277-1298.
40. Basso, J. C., McHale, A., Ende, V., Oberlin, D. J., & Suzuki, W. A. (2019). Brief, daily meditation enhances attention, memory, mood, and emotional regulation in non-experienced meditators. *Behavioural Brain Research, 356,* 208-220.
41. Whitfield, T., Barnhofer, T., Acabchuk, R., Cohen, A., Lee, M., Schlosser, M.,...Marchant, N. L. (2022). The effect of mindfulness-based programs on cognitive function in adults: A systematic review and meta-analysis. *Neuropsychology Review, 32*(3), 677-702.
42. Takimoto-Ohnishi, E., Ohnishi, J., & Murakami, K. (2012). Mind-body medicine: Effect of the mind on gene expression. *Personalized Medicine Universe, 1*(1), 2-6.
43. Mannekote Thippaiah, S., Pradhan, B., Voyiaziakis, E., Shetty, R., Iyengar, S., Olson, C., & Tang, Y. Y. (2022). Possible role of parvalbumin interneurons in meditation and psychiatric illness. *Journal of Neuropsychiatry and Clinical Neurosciences, 34*(2), 113-123.
44. McEwen, B. S. (2007). Physiology and neurobiology of stress and adaptation: Central role of the brain. *Physiological Reviews, 87*(3), 873-904.
45. Lupien, S. J., Parent, S., Evans, A. C., Tremblay, R. E., Zelazo, P. D., Corbo, V.,...Séguin, J. R. (2011). Larger amygdala but no change in hippocampal volume in 10-year-old children exposed to maternal depressive symptomatology since birth. *Proceedings of the National Academy of Sciences, 108*(34), 14324-14329.
46. McEwen, B. S., Nasca, C., & Gray, J. D. (2016). Stress effects on neuronal structure: Hippocampus, amygdala, and prefrontal cortex. *Neuropsychopharmacology, 41*(1), 3-23.

47. Davidson, R. J., & McEwen, B. S. (2012). Social influences on neuroplasticity: Stress and interventions to promote well-being. *Nature Neuroscience, 15*(5), 689-695.
48. McEwen, B. S. (2010). Stress, sex, and neural adaptation to a changing environment: Mechanisms of neuronal remodeling. *Annals of the New York Academy of Sciences, 1204,* 38-59.
49. Hölzel, B. K., Carmody, J., Evans, K. C., Hoge, E. A., Dusek, J. A., Morgan, L.,...Lazar, S. W. (2010). Stress reduction correlates with structural changes in the amygdala. *Social Cognitive and Affective Neuroscience, 5*(1), 11-17.
50. Karten, Y. J., Olariu, A., & Cameron, H. A. (2005). Stress in early life inhibits neurogenesis in adulthood. *Trends in Neurosciences, 28*(4), 171-172.
51. Chandramohan, Y., Droste, S. K., Arthur, J. S. C., & Reul, J. M. (2008). The forced swimming-induced behavioural immobility response involves histone H3 phospho acetylation and c-Fos induction in dentate gyrus granule neurons via activation of the N-methyl-d-aspartate/extracellular signal-regulated kinase/mitogen-and stress-activated kinase signalling pathway. *European Journal of Neuroscience, 27*(10), 2701-2713.
52. Ganesan, S., Moffat, B. A., Van Dam, N. T., Lorenzetti, V., & Zalesky, A. (2023). Meditation attenuates default-mode activity: A pilot study using ultra-high field 7 Tesla MRI. *Brain Research Bulletin, 203,* 110766.
53. Khalaf, O., Resch, S., Dixsaut, L., Gorden, V., Glauser, L., & Gräff, J. (2018). Reactivation of recall-induced neurons contributes to remote fear memory attenuation. *Science, 360*(6394), 1239-1242.
54. Seo, J., Cho, H., Kim, G. T., Kim, C. H., & Kim, D. G. (2017). Glutamatergic stimulation of the left dentate gyrus abolishes depressive-like behaviors in a rat learned helplessness paradigm. *Neuroimage, 159,* 207-213.
55. Botterill, J. J., Vinod, K. Y., Gerencer, K. J., Teixeira, C. M., LaFrancois, J. J., & Scharfman, H. E. (2021). Bidirectional regulation of cognitive and anxiety-like behaviors by dentate gyrus mossy cells in male and female mice. *Journal of Neuroscience, 41*(11), 2475-2495.
56. Statista. (2024). *Share of U.S. adults who said they had a family member or friend who had Alzheimer's disease as of 2023.* Retrieved from https://www.statista.com/statistics/1440198/adults-who-knew-someone-with-alzheimers-disease
57. Centers for Disease Control and Prevention (CDC). (2024). *Caregiving for a person with Alzheimer's disease or a related dementia.* Retrieved from https://www.cdc.gov/aging/caregiving/alzheimer.htm
58. Dwivedi, M., Dubey, N., Pansari, A. J., Bapi, R. S., Das, M., Guha, M.,...Ghosh, A. (2021). Effects of meditation on structural changes of the brain in patients with mild cognitive impairment or Alzheimer's disease dementia. *Frontiers in Human Neuroscience, 15,* 728993.
59. Zhao H., Li X., Wu W., Li Z., Qian L., Li S., et al. (2015). Atrophic patterns of the frontal-subcortical circuits in patients with mild cognitive impairment and Alzheimer's disease. *PLoS One, 10*(6), e0130017.
60. Newberg, A., & Waldman, M. R. (2017). *How enlightenment changes your brain: The new science of transformation.* Penguin.
61. Hosseini, S., Chaurasia, A., & Oremus, M. (2019). The effect of religion and spirituality on cognitive function: A systematic review. *Gerontologist, 59*(2), e76-e85.
62. Dos Santos, S. B., Rocha, G. P., Fernandez, L. L., De Padua, A. C., & Reppold, C. T. (2018). Association of lower spiritual well-being, social support, self-esteem, subjective well-being, optimism and hope scores with mild cognitive impairment and mild dementia. *Frontiers in Psychology, 9,* 371.
63. Khalsa, D. S., & Newberg, A. B. (2021). Spiritual fitness: a new dimension in Alzheimer's disease prevention. *Journal of Alzheimer's Disease, 80*(2), 505-519.
64. Amen, D. G., Krishnamani, P., Meysami, S., Newberg, A., & Raji, C. A. (2017). Classification of depression, cognitive disorders, and co-morbid depression and cognitive disorders with perfusion SPECT neuroimaging. *Journal of Alzheimer's Disease, 57*(1), 253-266.
65. Simons, M., Levin, J., & Dichgans, M. (2023). Tipping points in neurodegeneration. *Neuron, 111*(19), 2954-2968.
66. Goleman, D., & Davidson, R. J. (2018). *Altered traits: Science reveals how meditation changes your mind, brain, and body* (p. 78). Avery.

67. Ganesan, S., Moffat, B. A., Van Dam, N. T., Lorenzetti, V., & Zalesky, A. (2023). Meditation attenuates default-mode activity: A pilot study using ultra-high field 7 Tesla MRI. *Brain Research Bulletin, 203,* 110766.

Chapter 3

1. Perkins, H. (2021). *The muscle tipping point.* Retrieved from https://www.hollyperkins.com/blog/the-muscle-tipping-point%2F13231
2. Church, D. (2018). *Mind to matter: The astonishing science of how your brain creates material reality.* Hay House.
3. Kotler, S., & Wheal, J. (2017). *Stealing fire: How Silicon Valley, the Navy SEALs, and maverick scientists are revolutionizing the way we live and work.* HarperCollins.
4. Goleman, D., & Davidson, R. J. (2018). *Altered traits: Science reveals how meditation changes your mind, brain, and body.* Penguin.
5. Katsumi, Y., Kondo, N., Dolcos, S., Dolcos, F., & Tsukiura, T. (2021). Intrinsic functional network contributions to the relationship between trait empathy and subjective happiness. *NeuroImage, 227,* 117650.
6. Tanzer, J. R., & Weyandt, L. (2020). Imaging happiness: Meta analysis and review. *Journal of Happiness Studies, 21,* 2693-2734. Retrieved from https://www.sciencedirect.com/science/article/pii/S1053811920311356
7. Luo, Y., Kong, F., Qi, S., You, X., & Huang, X. (2016). Resting-state functional connectivity of the default mode network associated with happiness. *Social Cognitive and Affective Neuroscience, 11*(3), 516-524.
8. Johansson, M., Stomrud, E., Johansson, P. M., Svenningsson, A., Palmqvist, S., Janelidze, S.,...Hansson, O. (2022). Development of apathy, anxiety, and depression in cognitively unimpaired older adults: Effects of Alzheimer's disease pathology and cognitive decline. *Biological Psychiatry, 92*(1), 34-43.
9. Cunningham, W. A., & Kirkland, T. (2014). The joyful, yet balanced, amygdala: Moderated responses to positive but not negative stimuli in trait happiness. *Social Cognitive and Affective Neuroscience, 9*(6), 760-766.
10. Ran, Q., Yang, J., Yang, W., Wei, D., Qiu, J., & Zhang, D. (2017). The association between resting functional connectivity and dispositional optimism. *PLoS One, 12*(7), e0180334.
11. Miller, L. (2021). *The awakened brain: The new science of spirituality and our quest for an inspired life.* Random House.
12. Miller, L. (2013). Spiritual awakening and depression in adolescents: A unified pathway or "two sides of the same coin." *Bulletin of the Menninger Clinic, 77*(4), 332-348.
13. Miller, L., Bansal, R., Wickramaratne, P., Hao, X., Tenke, C. E., Weissman, M. M., & Peterson, B. S. (2014). Neuroanatomical correlates of religiosity and spirituality: A study in adults at high and low familial risk for depression. *JAMA Psychiatry, 71*(2), 128-135.
14. Miller, L. (2021). *The awakened brain: The new science of spirituality and our quest for an inspired life* (p. 237). Random House.
15. Frankl, V. (2017). *Man's search for meaning.* Beacon Press.
16. O'Hearn, B. (2015). Etty Hillesum. *Western Mystics.* Retrieved from https://westernmystics.wordpress.com/2015/06/18/etty-hillesum
17. Hillesum, E. (1996). *An interrupted life.* Picador.
18. Church, D. (1994). *Facing death, finding love: The healing power of grief and loss in one family's life.* Aslan.
19. Church, D. (2020). *Bliss brain: The neuroscience of remodeling your brain for resilience, creativity, and joy.* Hay House.
20. Martin, J. A. (2019). *The finders.* Integration Press.
21. Martin, J. A., Ericson, M., Berwaldt, A., Stephens, E. D., & Briner, L. (2021). Effects of two online positive psychology and meditation programs on persistent self-transcendence. *Psychology of Consciousness: Theory, Research, and Practice, 10*(3), 225.
22. Kotler, S., & Wheal, J. (2017). *Stealing fire: How Silicon Valley, the Navy SEALs, and maverick scientists are revolutionizing the way we live and work.* HarperCollins.
23. Martin, J. A. (2019). *The finders* (p. 16). Integration Press.

24. Martin, J. A. (2019). *The finders* (p. 32). Integration Press.
25. Ericsson, K. A., Hoffman, R. R., Kozbelt, A., & Williams, A. M. (Eds.). (2018). *The Cambridge handbook of expertise and expert performance*. Cambridge University Press.
26. Ericsson, K. A., & Lehmann, A. C. (1996). Expert and exceptional performance: Evidence of maximal adaptation to task constraints. *Annual Review of Psychology, 47*(1), 273.
27. Selye, H. (1956). *The stress of life*. McGraw-Hill.
28. Benson, H. (1975). *The relaxation response*. Morrow.
29. Pew Research Center. (2018). *Being Christian in Western Europe*. Retrieved from https://www.pewresearch.org/religion/2018/05/29/being-christian-in-western-europe
30. Pew Research Center. (2023). *How the pandemic has affected attendance at U.S. religious services*. Retrieved from https://www.pewresearch.org/religion/2023/03/28/how-the-pandemic-has-affected-attendance-at-u-s-religious-services
31. World Health Organization. (2022). *Mental health and COVID-19: Early evidence of the pandemic's impact: Scientific brief, 2 March 2022* (No. WHO/2019-nCoV/Sci_Brief/Mental_health/2022.1). World Health Organization.
32. Brunton, P. (1969). *The secret path*. Rider.
33. Brunton, P. (1934). *A search in secret India*. Rider.
34. Brunton, P. (2014). *The short path to enlightenment: Instructions for immediate awakening*. Paul Brunton Philosophic Foundation/Larson Publications.
35. *Ibid.*
36. *Ibid.*
37. Stapleton, P., Church, D., Baumann, O., & Sabot, D. (2022). EcoMeditation modifies brain resting state network activity. *Innovations in Clinical Neuroscience, 19*(7-9), 61-70.
38. Pennington, J., Sabot, D., & Church, D. (2019). EcoMeditation and Emotional Freedom Techniques (EFT) produce elevated brain-wave patterns and states of consciousness. *Energy Psychology: Theory, Research, and Treatment, 11*(1), 13-40.
39. Church, D., Stapleton, P., & Sabot, D. (2020). Brief EcoMeditation associated with psychological improvements: A preliminary study. *Global Advances in Health and Medicine, 9.*

Chapter 4

1. Roman, J. (2019). Re-thinking Thoreau: Between the lines of his life and work. *Bangalore Review*. Retrieved from https://bangalorereview.com/2019/12/re-thinking-thoreau-between-the-lines-of-his-life-and-work
2. Martin, J. A. (2019). *The finders*. Integration Press.
3. Goleman, D., & Davidson, R. J. (2018). *Altered traits: Science reveals how meditation changes your mind, brain, and body*. Penguin.
4. Greeley, A. (1974). *Ecstasy: A way of knowing*. Prentice Hall.
5. Kotler, S., & Wheal, J. (2017). *Stealing fire: How Silicon Valley, the Navy SEALs, and maverick scientists are revolutionizing the way we live and work*. HarperCollins.
6. McClenon, J. (1993). Surveys of anomalous experience in Chinese, Japanese, and American samples. *Sociology of Religion, 54*(3), 295-302.
7. Church, D. (2020). *Bliss brain: The neuroscience of remodeling your brain for resilience, creativity, and joy*. Hay House.
8. Newberg, A., & Waldman, M. R. (2017). *How enlightenment changes your brain: The new science of transformation*. Penguin.
9. Goleman, D., & Davidson, R. J. (2018). *Altered traits: Science reveals how meditation changes your mind, brain, and body*. Penguin.
10. Newberg, A., & Waldman, M. R. (2017). *How enlightenment changes your brain: The new science of transformation*. Penguin.
11. Church, D., Vasudevan, A., & De Foe, A. (2024). The association of relational spirituality in an EcoMeditation course with flow, transcendent states, and professional productivity. *Advances in Mind Body Medicine, 28*(3), 4-14.
12. Church, D., Stapleton, P., Gosatti, D., & O'Keefe, T. (2022). Effect of virtual group EcoMeditation on psychological conditions and flow states. *Frontiers in Psychology, 13*, 907846.
13. Martin, J. A. (2019). *The finders*. Integration Press.

14. Martin, J. A., Ericson, M., Berwaldt, A., Stephens, E. D., & Briner, L. (2023). Effects of two online positive psychology and meditation programs on persistent self-transcendence. *Psychology of Consciousness: Theory, Research, and Practice, 10*(3), 225.
15. Martin, J. A. (2019). *The finders.* Integration Press.
16. Church, D. (2021). The finders—Part 1: Jeffery Martin and Dawson Church in conversation. *High Energy Health Podcast.* Retrieved from https://highenergyhealthpodcast.podbean.com/e/the-finders-part-1-jeffery-martin-and-dawson-church-in-conversation
17. Morris, M. (2023). *8 secrets to powerful manifesting: How to create the reality of your dreams.* Hay House.
18. Brewer, J. A., Worhunsky, P. D., Gray, J. R., Tang, Y. Y., Weber, J., & Kober, H. (2011). Meditation experience is associated with differences in default mode network activity and connectivity. *Proceedings of the National Academy of Sciences, 108*(50), 20254-20259.
19. Osborne, A. (2013). *The teachings of Sri Ramana Maharshi in his own words.* Sri Ramanasramam.
20. Tigunait, P. R. (2017). *The secret of the Yoga Sutra: Samadhi Pada.* Audiobook. Himalayan Institute.
21. Viveka-Chudamani. (1970). *Shankara's crest jewel of discrimination.* Translated by Swami Prabhavananda & C. Isherwood. Vedanta.
22. Gabriel, R. (2019). *The 3 levels of Samadhi.* Retrieved from https://chopra.com/blogs/meditation/the-3-levels-of-samadhi
23. Satchidananda. (n.d.). *The stages of Samadhi.* Retrieved from https://integralyogamagazine.org/samadhi-the-different-stages
24. Stapleton, P., Church, D., Baumann, O., & Sabot, D. (2022). EcoMeditation modifies brain resting state network activity. *Innovations in Clinical Neuroscience, 19*(7-9), 61-70.
25. *Ibid.*
26. Brunton, P. (1934). *A search in secret India.* Rider.
27. Boccia, M., Piccardi, L., & Guariglia, P. (2015). The meditative mind: A comprehensive meta-analysis of MRI studies. *BioMed Research International, 2015,* 419808.

Chapter 5

1. Wise, A. (1997). *The high-performance mind: Mastering brainwaves for insight, healing, and creativity.* Penguin.
2. Cade, M. & Coxhead, N. (1979). *The awakened mind: Biofeedback and the development of higher states of awareness.* Dell.
3. Braboszcz, C., Cahn, B. R., Levy, J., Fernandez, M., & Delorme, A. (2017). Increased gamma brainwave amplitude compared to control in three different meditation traditions. *PloS One, 12*(1), e0170647.
4. Goleman, D., & Davidson, R. J. (2018). *Altered traits: Science reveals how meditation changes your mind, brain, and body.* Penguin.
5. Lee, D. J., Kulubya, E., Goldin, P., Goodarzi, A., & Girgis, F. (2018). Review of the neural oscillations underlying meditation. *Frontiers in Neuroscience, 12,* 320145.
6. Groesbeck, G., Bach, D., Stapleton, P., Blickheuser, K., Church, D., & Sims, R. (2018). The interrelated physiological and psychological effects of EcoMeditation. *Journal of Evidence-Based Integrative Medicine, 23.*
7. Cade, M. & Coxhead, N. (1979). *The awakened mind: Biofeedback and the development of higher states of awareness.* Dell.
8. Strathern, P. (2000). *Mendeleyev's dream: The quest for the elements.* St. Martin's Press.
9. Csikszentmihalyi, M. (1990). *Flow: The psychology of optimal experience.* Harper & Row.
10. Patillo, N. (2017). The 'flow' of peak performance. *CNN.* Retrieved from https://www.cnn.com/2017/08/02/health/flow-peak-experience/index.html
11. Maslow, A. H. (1971). *Self-actualization.* Big Sur Recordings.
12. Koltko-Rivera, M. E. (2006). Rediscovering the later version of Maslow's hierarchy of needs: Self-transcendence and opportunities for theory, research, and unification. *Review of General Psychology, 10*(4), 302-317.
13. Maslow, Abraham H. (1996). Critique of self-actualization theory. In E. Hoffman (Ed.), *Future visions: The unpublished papers of Abraham Maslow* (pp. 26-32). Sage.
14. Csikszentmihalyi, M. (1990). *Flow: The psychology of optimal experience.* Harper Perennial.

15. Pappas, A. (2022). *Bravey: Chasing dreams, befriending pain, and other big ideas.* Dial.
16. Litvan, K. (2024). Beyond the finish line: Olympian Alexi Pappas shares stories, advice and insights. *Elon University.* Retrieved from https://www.elon.edu/u/news/2024/03/13/beyond-the-finish-line-olympian-alexi-pappas-shares-stories-advice-and-insights
17. *Ibid.*
18. Volpe, A. (2019). What a runner's high actually feels like according to runners. *Medium.* Retrieved from https://elemental.medium.com/what-a-runners-high-actually-feels-like-5bbc887a071d
19. Church, D., Stapleton, P., Yang, A., & Gallo, F. (2018). Is tapping on acupuncture points an active ingredient in Emotional Freedom Techniques (EFT)? A systematic review and meta-analysis of comparative studies. *Journal of Nervous and Mental Disease,* 206(10), 783-793.
20. Church, D., Yount, G., & Brooks, A. J. (2012). The effect of Emotional Freedom Techniques on stress biochemistry: A randomized controlled trial. *Journal of Nervous and Mental Disease, 200*(10), 891-896.
21. Schmidt, J. E., Carlson, C. R., Usery, A. R., & Quevedo, A. S. (2009). Effects of tongue position on mandibular muscle activity and heart rate function. *Oral Surgery, Oral Medicine, Oral Pathology, Oral Radiology, and Endodontology, 108*(6), 881-888.
22. Fehmi, L., & Robbins, J. (2008). *The open-focus brain: Harnessing the power of attention to heal mind and body.* Shambhala/Trumpeter Books.
23. McCraty, R., Atkinson, M., Tomasino, D., & Bradley, R. T. (2009). The coherent heart: Heart-brain interactions, psychophysiological coherence, and the emergence of system-wide order. *Integral Review: A Transdisciplinary and Transcultural Journal for New Thought, Research, and Praxis, 5*(2), 7-14.
24. *Ibid.*
25. McCraty, R. (2003). *The energetic heart: Bioelectromagnetic interactions within and between people.* Institute of HeartMath.
26. Siegel, D. J. (2022). *IntraConnected: MWe (Me+ We) as the integration of self, identity, and belonging* (Norton Series on Interpersonal Neurobiology). W. W. Norton.
27. Newberg, A., & Waldman, M. R. (2017). *How enlightenment changes your brain: The new science of transformation.* Penguin.
28. Wheal, J. (2021). *Recapture the rapture: Rethinking God, sex, and death in a world that's lost its mind.* Harper.
29. *Ibid.*
30. Kotler, S. (2021). *The art of impossible: A peak performance primer.* HarperCollins.
31. *Ibid.*
32. Newberg, A., & Waldman, M. R. (2017). *How enlightenment changes your brain: The new science of transformation.* Penguin.
33. Pennington, J., Sabot, D., & Church, D. (2019). EcoMeditation and Emotional Freedom Techniques (EFT) produce elevated brain-wave patterns and states of consciousness. *Energy Psychology: Theory, Research, and Treatment, 11*(1), 13-40.
34. Van Der Linden, D., Tops, M., & Bakker, A. B. (2021). Go with the flow: A neuroscientific view on being fully engaged. *European Journal of Neuroscience, 53*(4), 947-963.
35. Stapleton, P., Church, D., Baumann, O., & Sabot, D. (2022). EcoMeditation modifies brain resting state network activity. *Innovations in Clinical Neuroscience, 19*(7-9), 61-70.
36. Kotler, S. (2020). *What are the brain waves of flow?* Retrieved from https://www.facebook.com/watch/?v=3081469098601137
37. Beaty, R. (2018). Study reveals why some people are more creative than others. *NeuroscienceNews.* Retrieved from https://neurosciencenews.com/neuroscience-creativity-8325
38. Kotler, S. (2023). How to get into flow state: 8 tips to unlock peak performance. *Flow Research Collective.* Retrieved from https://www.flowresearchcollective.com/blog/how-to-get-into-flow-state
39. Church, D., Vasudevan, A., & De Foe, A. (2024). The association of relational spirituality in an EcoMeditation course with flow, transcendent states, and professional productivity. *Advances in Mind-Body Medicine, 28*(3), 4-14.
40. Amram, Y. (2023). *Spiritually intelligent leadership: How to inspire by being inspired.* Waterside.

41. Amram, J. Y. (2009). *The contribution of emotional and spiritual intelligences to effective business leadership.* ITP Dissertation. Retrieved from https://media.proquest.com/media/hms/ORIG/2/kylxI?_s=ww5Fc%2B7Mp%2BPYuQ%2B4TZUsFxyUBRY%3D
42. Ayranci, E. (2011). Effects of top Turkish managers' emotional and spiritual intelligences on their organizations' financial performance. *Business Intelligence Journal, 4*(1), 9-32.
43. Malik, M. S., & Tariq, S. (2016). Impact of spiritual intelligence on organizational performance. *International Review of Management and Marketing, 6*(2), 289-297.
44. Yahyazadeh-Jeloudar, S., & Lotfi-Goodarzi, F. (2012). What is the relationship between spiritual intelligence and job satisfaction among MA and BA teachers? *International Journal of Business and Social Science, 3*(8), 299-303.
45. Koražija, M., Žižek, S. Š., & Mumel, D. (2016). The relationship between spiritual intelligence and work satisfaction among leaders and employees. *Naše Gospodarstvo/Our Economy, 62*(2), 51-60.
46. Rani, A. A., Abidin, I., & Hamid, M. R. (2013). The impact of spiritual intelligence on work performance: Case studies in government hospitals of east coast of Malaysia. *Macrotheme Review, 2*(3), 46-59.
47. Liu, Z., Li, X., Jin, T., Xiao, Q., & Wuyun, T. (2021). Effects of ethnicity and spiritual intelligence in the relationship between awe and life satisfaction among Chinese primary school teachers. *Frontiers in Psychology, 12,* 673832.
48. Narayanan, A., & Jose, T. P. (2011). Spiritual intelligence and resilience among Christian youth in Kerala. *Journal of the Indian Academy of Applied Psychology, 37*(2), 263-268.
49. Harung, H. S., & Travis, F. (2012). Higher mind-brain development in successful leaders: Testing a unified theory of performance. *Cognitive Processing, 13,* 171-181.
50. Cranston, S., & Keller, S. (2013). Increasing the meaning quotient of work. *McKinsey Quarterly, 1,* 48-59.
51. Adee, S. (2012). Zap your brain into the zone: Fast track to pure focus. *New Scientist, 2850,* 1-6.
52. Amabile, T. M., & Pillemer, J. (2012). Perspectives on the social psychology of creativity. *Journal of Creative Behavior, 46*(1), 3-15.
53. Soga, L. R., Bolade-Ogunfodun, Y., Mariani, M., Nasr, R., & Laker, B. (2022). Unmasking the other face of flexible working practices: A systematic literature review. *Journal of Business Research, 142,* 648-662.
54. Wheal, J. (2021). *Recapture the rapture: Rethinking God, sex, and death in a world that's lost its mind.* Harper.
55. Zohar, D. (2015). Spiritual intelligence and business success. *Bangkok Post.* Retrieved from https://www.bangkokpost.com/business/456975/spiritual-intelligence-and-business-success
56. Phillips, S., & Mychailyszyn, M. (2022). The effect of school-based mindfulness interventions on anxious and depressive symptoms: A meta-analysis. *School Mental Health, 14*(3), 455-469.
57. Miller, L. (2021). *The awakened brain: The new science of spirituality and our quest for an inspired life.* Random House.
58. Narayanan, A., & Jose, T. P. (2011). Spiritual intelligence and resilience among Christian youth in Kerala. *Journal of the Indian Academy of Applied Psychology, 37*(2), 263-268.
59. Han, A. (2022). Effects of mindfulness-based interventions on psychological distress and mindfulness in incarcerated populations: A systematic review and meta-analysis. *Criminal Behaviour and Mental Health, 32*(1), 48-59.

Chapter 6

1. Darwin, C. (1859). *On the origin of species by means of natural selection, or the preservation of favoured races in the struggle for life* (p. 84). John Murray.
2. *Ibid.*
3. Mayr, E. (2001). *What evolution is.* Basic Books.
4. Stegner, M. E., Stemme, T., Iliffe, T. M., Richter, S., & Wirkner, C. S. (2015). The brain in three crustaceans from cavernous darkness. *BMC Neuroscience, 16,* 1-29.
5. Hullinger, J. (2021). 6 animals that are rapidly evolving. *Mental Floss.* Retrieved from https://www.mentalfloss.com/article/64300/6-animals-are-rapidly-evolving

6. Eldredge, N., & Gould, S. J. (1972). Punctuated equilibria: An alternative to phyletic gradualism. In T. J. M. Schopf (Ed.), *Models in paleobiology* (pp. 82-115). Freeman, Cooper.
7. Venditti, C., & Pagel, M. (2008). Speciation and bursts of evolution. *Evolution: Education and Outreach, 1,* 274-280.
8. Gould, S. J., & Eldredge, N. (1977). Punctuated equilibria: The tempo and mode of evolution reconsidered. *Paleobiology, 3*(2), 115.
9. Vasquez, A., Alaniz, A., Dearth, R., & Kariyat, R. (2024). Continuous mowing differentially affects floral defenses in the noxious and invasive weed Solanum elaeagnifolium in its native range. *Scientific Reports, 14*(1), 8133.
10. Maron, J. L., Johnson, M. T., Hastings, A. P., & Agrawal, A. A. (2018). Fitness consequences of occasional outcrossing in a functionally asexual plant *(Oenothera biennis). Ecology, 99*(2), 464-473.
11. Burke, J. M., Knapp, S. J., & Rieseberg, L. H. (2002). Genetic consequences of selection during the evolution of cultivated sunflower. *Genetics, 161*(3), 1257-1267.
12. Franks, S. J., Sim, S., & Weis, A. E. (2007). Rapid evolution of flowering time by an annual plant in response to a climate fluctuation. *Proceedings of the National Academy of Sciences, 104*(4), 1278-1282.
13. Cabin, Z., Derieg, N. J., Garton, A., Ngo, T., Quezada, A., Gasseholm, C.,...Hodges, S. A. (2022). Non-pollinator selection for a floral homeotic mutant conferring loss of nectar reward in Aquilegia coerulea. *Current Biology, 32*(6), 1332-1341.
14. University of California–Santa Barbara (UCSB). (2022) Sudden evolutionary change in flowers. *Science Daily.* Retrieved from https://www.sciencedaily.com/releases/2022/02/220216112300.htm
15. Campbell-Staton, S. C., Arnold, B. J., Gonçalves, D., Granli, P., Poole, J., Long, R. A., & Pringle, R. M. (2021). Ivory poaching and the rapid evolution of tusklessness in African elephants. *Science, 374*(6566), 483-487.
16. Rensselaer Polytechnic Institute (RPI). (2024). Frog species evolved rapidly in response to road salts. *Science Daily.* Retrieved from https://www.sciencedaily.com/releases/2024/04/240423184759.htm?
17. Stuart, Y. E., Campbell, T. S., Hohenlohe, P. A., Reynolds, R. G., Revell, L. J., & Losos, J. B. (2014). Rapid evolution of a native species following invasion by a congener. *Science, 346*(6208), 463-466.
18. Hullinger, J. (2021). 6 animals that are rapidly evolving. Mental Floss. Retrieved from https://www.mentalfloss.com/article/64300/6-animals-are-rapidly-evolving
19. Cucherousset, J., Azemar, F., Compin, A., Guillaume, M., Santoul, F., & Roussel, J. M. (2012). 'Freshwater killer whales': Beaching behavior of an alien fish to hunt land birds. *PLoS One, 7*(12), e50840.
20. Baym, M., Lieberman, T. D., Kelsic, E. D., Chait, R., Gross, R., Yelin, I., & Kishony, R. (2016). Spatiotemporal microbial evolution on antibiotic landscapes. *Science, 353*(6304), 1147-1151.
21. Yi, X., Liang, Y., Huerta-Sanchez, E., Jin, X., Cuo, Z. X. P., Pool, J. E., . . . Wang, J. (2010). Sequencing of 50 human exomes reveals adaptation to high altitude. *Science, 329*(5987), 75-78.
22. Sanders, R. (2010). Tibetans adapted to high altitude in less than 3,000 years. *Berkeley News.* Retrieved from https://news.berkeley.edu/2010/07/01/tibetan_genome
23. Timmann, C., Thye, T., Vens, M., Evans, J., May, J., Ehmen, C.,...Horstmann, R. D. (2012). Genome-wide association study indicates two novel resistance loci for severe malaria. *Nature, 489*(7416), 443-446.
24. Hollox, E. J., & Hoh, B. P. (2014). Human gene adaptation to malaria and other infectious diseases. Human Genetics, 133(3), 327-334.
25. Gibbons, A. (2014). The evolution of diet. *National Geographic, 226*(3), 34-53.
26. Itan, Y., Powell, A., Beaumont, M. A., Burger, J., & Thomas, M. G. (2009). The origins of lactase persistence in Europe. *PLoS Computational Biology, 5*(8), e1000491.
27. Schwartz, G. T. (2007). Dental development and the evolution of human life history. *Evolutionary Anthropology: Issues, News, and Reviews, 16*(4), 205-213.

28. Kaliman, P., Alvarez-Lopez, M. J., Cosín-Tomás, M., Rosenkranz, M. A., Lutz, A., & Davidson, R. J. (2014). Rapid changes in histone deacetylases and inflammatory gene expression in expert meditators. *Psychoneuroendocrinology, 40,* 96-107.
29. Kaliman, P. (2019). Epigenetics and meditation. *Current Opinion in Psychology, 28,* 76-80.
30. Heijmans, B. T., Tobi, E. W., Stein, A. D., Putter, H., Blauw, G. J., Susser, E. S.,...Lumey, L. H. (2008). Persistent epigenetic differences associated with prenatal exposure to famine in humans. *Proceedings of the National Academy of Sciences, 105*(44), 17046-17049.
31. Oberlander, T. F., Weinberg, J., Papsdorf, M., Grunau, R., Misri, S., & Devlin, A. M. (2008). Prenatal exposure to maternal depression, neonatal methylation of human glucocorticoid receptor gene (NR3C1) and infant cortisol stress responses. *Epigenetics, 3*(2), 97-106.
32. Hoffmann, L. B., Li, B., Zhao, Q., Wei, W., Leighton, L. J., Bredy, T. W.,...Hannan, A. J. (2024). Chronically high stress hormone levels dysregulate sperm long noncoding RNAs and their embryonic microinjection alters development and affective behaviours. *Molecular Psychiatry, 29*(3), 590-601.
33. Joubert, B. R., Haberg, S. E., Nilsen, R. M., Wang, X., Vollset, S. E., Murphy, S. K.,...London, S. J. (2016). 450K epigenome-wide scan identifies differential DNA methylation in newborns related to maternal smoking during pregnancy. *Environmental Health Perspectives, 124*(10), 1444-1453.
34. Yehuda, R., Daskalakis, N. P., Bierer, L. M., Bader, H. N., Klengel, T., Holsboer, F., & Binder, E. B. (2016). Holocaust exposure induced intergenerational effects on FKBP5 methylation. *Biological Psychiatry, 80*(5), 372-380.
35. Arkowitz, H., & Lilienfeld, S. O. (2007). Why don't people change? *Scientific American.* Retrieved from https://www.scientificamerican.com/article/why-dont-people-change
36. Robinson, M. D., Irvin, R. L., Pringle, T. A., & Klein, R. J. (2023). General cognitive ability, as assessed by self-reported ACT scores, is associated with reduced emotional responding: Evidence from a Dynamic Affect Reactivity Task. *Intelligence, 99,* 101760.
37. Lenzi, S. C., Cossell, L., Grainger, B., Olesen, S. F., Branco, T., & Margrie, T. W. (2022). Threat history controls flexible escape behavior in mice. *Current Biology, 32*(13), 2972-2979.
38. Sainsbury Wellcome Centre. (2022). New study shows mice robustly learn to suppress their innate escape responses. *Science Daily.* Retrieved from https://www.sciencedaily.com/releases/2022/06/220602114217.htm
39. Church, D. (2020). *Bliss brain: The neuroscience of remodeling your brain for resilience, creativity, and joy.* Hay House.
40. Szeska, C., Pünjer, H., Riemann, S., Meinzer, M., & Hamm, A. O. (2022). Stimulation of the ventromedial prefrontal cortex blocks the return of subcortically mediated fear responses. *Translational Psychiatry, 12,* 394.
41. Hedrih, V. (2022). Electrical stimulation to a specific region of the brain can block the return of fear responses, study finds. *Psypost.* Retrieved from https://www.psypost.org/electrical-stimulation-to-a-specific-region-of-the-brain-can-block-the-return-of-fear-responses-study-finds
42. Martin, J. A. (2019). *The finders.* Integration Press.
43. Paige, J., & Perreault, C. (2024). 3.3 million years of stone tool complexity suggests that cumulative culture began during the Middle Pleistocene. *Proceedings of the National Academy of Sciences, 121*(26), e2319175121.
44. Arizona State University (ASU). (2024). Origins of cumulative culture in human evolution. *Science Daily.* Retrieved from https://www.sciencedaily.com/releases/2024/06/240617173730.htm?
45. *Ibid.*
46. Waring, T. M., & Wood, Z. T. (2021). Long-term gene–culture coevolution and the human evolutionary transition. *Proceedings of the Royal Society B, 288*(1952), 20210538.
47. University of Maine. (2021). Culture drives human evolution more than genetics. *Science Daily.* Retrieved from https://www.sciencedaily.com/releases/2021/06/210602170624.htm
48. Decety, J., & Jackson, P. L. (2004). The functional architecture of human empathy. *Behavioral and Cognitive Neuroscience Reviews, 3*(2), 71-100.
49. Church, D. (2020). *Bliss brain: The neuroscience of remodeling your brain for resilience, creativity, and joy.* Hay House.

50. Leetaru, K. (2011). Culturomics 2.0: Forecasting large-scale human behavior using global news media tone in time and space. *First Monday, 16*(9). doi:10.5210/fm.v16i9.3663
51. Darwin, C. (1871). *The descent of man, and selection in relation to sex.* D. Appleton.
52. Paulus, M., Becher, T., Christner, N., Kammermeier, M., Gniewosz, B., & Pletti, C. (2024). When do children begin to care for others? The ontogenetic growth of empathic concern across the first two years of life. *Cognitive Development, 70,* 101439.
53. Conde-Valverde, M., Quirós-Sánchez, A., Diez-Valero, J., Mata-Castro, N., García-Fernández, A., Quam, R.,...Villaverde, V. (2024). The child who lived: Down syndrome among Neanderthals? *Science Advances, 10*(26), eadn9310.
54. Goetz, J. L., Keltner, D.,& Simon-Thomas, E. (2010). Compassion: an evolutionary analysis and empirical review. *Psychological Bulletin, 136*(3), 351.
55. Stapleton, P., Church, D., Baumann, O., & Sabot, D. (2022). EcoMeditation modifies brain resting state network activity. *Innovations in Clinical Neuroscience, 19*(7–9), 61–70.
56. Davidson, R. J., & Lutz, A. (2008). Buddha's brain: Neuroplasticity and meditation. *IEEE Signal Processing Magazine, 25*(1), 176.
57. Church, D. (2020). *Bliss brain: The neuroscience of remodeling your brain for resilience, creativity, and joy.* Hay House.
58. Clarke, T. C., Barnes, P. M., Black, L. I., Stussman, B. J., & Nahin, R. L. (2018). *Use of yoga, meditation, and chiropractors among US adults aged 18 and over* (pp. 1-8). Washington, DC, USA: US Department of Health and Human Services, Centers for Disease Control and Prevention, National Center for Health Statistics.
59. Wright, M. J., Galante, J., Corneille, J. S., Grabovac, A., Ingram, D. M., & Sacchet, M. D. (2024). Altered states of consciousness are prevalent and insufficiently supported clinically: A population survey. *Mindfulness,* 1-14.
60. Brown, N. (2024). Yoga and meditation-induced altered states of consciousness are common in the general population. *Massachusetts General Hospital.* Retrieved from https://www.massgeneral.org/news/press-release/yoga-meditation-induced-altered-states-of-consciousness
61. Wright, M. J., Galante, J., Corneille, J. S., Grabovac, A., Ingram, D. M., & Sacchet, M. D. (2024). Altered states of consciousness are prevalent and insufficiently supported clinically: A population survey. *Mindfulness,* 1-14.
62. Vieten, C., Wahbeh, H., Cahn, B. R., MacLean, K., Estrada, M., Mills, P.,...Delorme, A. (2018). Future directions in meditation research: Recommendations for expanding the field of contemplative science. *PloS One, 13*(11), e0205740.
63. Jones, M. L. (2006). The growth of nonprofits. *Bridgewater Review, 25*(1), 13-17. Retrieved from https://vc.bridgew.edu/br_rev/vol25/iss1/8
64. Grapevine. (2023). The overcrowded nonprofit market. *Grapevine.* Retrieved from https://www.grapevine.org/blog/the-overcrowded-nonprofit-market
65. DeCarli, C., Maillard, P., Pase, M. P., Beiser, A. S., Kojis, D., Satizabal, C. L.,...Seshadri, S. (2024). Trends in intracranial and cerebral volumes of Framingham heart study participants born 1930 to 1970. *JAMA Neurology, 81*(5), 471-480.
66. Davis Health. (2024). Human brains are getting larger. *Eureka Alert.* Retrieved from https://www.eurekalert.org/news-releases/1039015
67. Luders, E., Toga, A. W., Lepore, N., & Gaser, C. (2009). The underlying anatomical correlates of long-term meditation: Larger hippocampal and frontal volumes of gray matter. *NeuroImage, 45*(3), 672-678.
68. Grinberg-Zylberbaum, J., Delaflor, M., Attie, L., & Goswami, A. (1994). The Einstein-Podolsky-Rosen paradox in the brain: The transferred potential. *Physics Essays, 7*(4), 422-428.
69. Newberg, A. (2018). *Neurotheology: How science can enlighten us about spirituality.* Columbia University Press.
70. Laszlo, E. (2007). *Science and the Akashic Record* (p. 143). Inner Traditions.
71. Lahey, S. (2023). Your very own consciousness can interact with the whole universe, scientists believe. *Popular Mechanics.* Retrieved from https://www.popularmechanics.com/science/a45574179/architecture-of-consciousness
72. McCraty, R., Atkinson, M., & Bradley, R. T. (2004). Electrophysiological evidence of intuition: Part 1. The surprising role of the heart. *Journal of Alternative and Complementary Medicine, 10*(1), 133-143.

73. McCraty, R., Atkinson, M., Tomasino, D., & Bradley, R. T. (2009). The coherent heart: Heart-brain interactions, psychophysiological coherence, and the emergence of system-wide order. *Integral Review: A Transdisciplinary and Transcultural Journal for New Thought, Research, and Praxis, 5*(2), 7-14.
74. Vazza, F., & Feletti, A. (2020). The quantitative comparison between the neuronal network and the cosmic web. *Frontiers in Physics, 8*, 525731.
75. Childers, T. (2020). The human brain looks suspiciously like the universe, which may freak you out. *Popular Mechanics*. Retrieved from https://www.popularmechanics.com/science/a34703841/human-brain-universe-similarities
76. Università di Bologna. (2020). Does the human brain resemble the Universe? *PhysOrg*. Retrieved from https://phys.org/news/2020-11-human-brain-resemble-universe.html
77. *Ibid.*
78. Laszlo, E. (2007). *Science and the Akashic Record*. Inner Traditions.
79. Church, D. (2018). *Mind to matter: The astonishing science of how your brain creates material reality*. Hay House.
80. Laszlo, E. (2007). *Science and the Akashic Record*. Inner Traditions.
81. *Ibid.*

Chapter 7

1. Pinker, S. (2019). *Enlightenment now: The case for reason, science, humanism, and progress* (p. 246). Penguin.
2. Rijpma, A. (2014). A composite view of well-being since 1820. In J. L. van Zanden, J. Baten, M. M. d'Ercole, A. Rijpma, & M. P. Timmer (eds.), *How was life?: Global well-being since 1820* (pp. 249-269). OECD Publishing. doi:10.1787/9789264214262-17-en
3. Prados de la Escosura, L. (2015). World human development: 1870–2007. *Review of Income and Wealth, 61*(2), 220-247.
4. Pinker, S. (2019). *Enlightenment now: The case for reason, science, humanism, and progress* (p. 246). Penguin.
5. *Ibid.*
6. Acerbi, A., & Brand, C. (2022). *Why are pop songs getting sadder than they used to be?* Retrieved from https://aeon.co/ideas/why-are-pop-songs-getting-sadder-than-they-used-to-be
7. Rozado, D., Hughes, R., & Halberstadt, J. (2022). Longitudinal analysis of sentiment and emotion in news media headlines using automated labelling with Transformer language models. *PloS One, 17*(10), e0276367.
8. Brooks, D. (2022). The rising tide of global sadness. *New York Times*. Retrieved from https://www.nytimes.com/2022/10/27/opinion/global-sadness-rising.html
9. Ray, J. (2022). *Who are the unhappiest people in the world?* Retrieved from https://news.gallup.com/opinion/gallup/400667/unhappiest-people-world.aspx
10. Martin, J. A. (2019). *The finders*. Integration Press.
11. Centers for Disease Control and Prevention (CDC). (2018). *Use of yoga and meditation becoming more popular in U.S.* Retrieved from https://www.cdc.gov/nchs/pressroom/nchs_press_releases/2018/201811_Yoga_Meditation.htm
12. Masci, D. & Hackett, C. (2018). Meditation is common across many religious groups in the U.S. *Pew Research Center*. Retrieved from https://www.pewresearch.org/short-reads/2018/01/02/meditation-is-common-across-many-religious-groups-in-the-u-s
13. Levine, G. N., Lange, R. A., Bairey-Merz, C. N., Davidson, R. J., Jamerson, K., Mehta, P. K.,...American Heart Association Council on Clinical Cardiology; Council on Cardiovascular and Stroke Nursing; and Council on Hypertension. (2017). Meditation and cardiovascular risk reduction: A scientific statement from the American Heart Association. *Journal of the American Heart Association, 6*(10), e002218.
14. Groesbeck, G., Bach, D., Stapleton, P., Blickheuser, K., Church, D., & Sims, R. (2018). The interrelated physiological and psychological effects of EcoMeditation. *Journal of Evidence-Based Integrative Medicine, 23.*

15. Bach, D., Groesbeck, G., Stapleton, P., Banton, S., Blickheuser, K., & Church, D. (2019). Clinical EFT (Emotional Freedom Techniques) improves multiple physiological markers of health. *Journal of Evidence-Based Integrative Medicine, 24.*
16. Black, D. S., & Slavich, G. M. (2016). Mindfulness meditation and the immune system: A systematic review of randomized controlled trials. *Annals of the New York Academy of Sciences, 1373*(1), 13-24.
17. Kaliman, P., Alvarez-Lopez, M. J., Cosín-Tomás, M., Rosenkranz, M. A., Lutz, A., & Davidson, R. J. (2014). Rapid changes in histone deacetylases and inflammatory gene expression in expert meditators. *Psychoneuroendocrinology, 40,* 96-107.
18. Del Val, C., Díaz de la Guardia-Bolívar, E., Zwir, I., Mishra, P. P., Mesa, A., Salas, R.,... Cloninger, C. R. (2024). Gene expression networks regulated by human personality. *Molecular Psychiatry,* 1-20.
19. Ornish, D., Madison, C., Kivipelto, M., Kemp, C., McCulloch, C. E., Galasko, D.,...Arnold, S. E. (2024). Effects of intensive lifestyle changes on the progression of mild cognitive impairment or early dementia due to Alzheimer's disease: A randomized, controlled clinical trial. *Alzheimer's Research and Therapy, 16*(1), 122.
20. Pinto, C. T., Guedes, L., Pinto, S., & Nunes, R. (2024). Spiritual intelligence: A scoping review on the gateway to mental health. *Global Health Action, 17*(1), 2362310.
21. Kaliman, P. (2019). Epigenetics and meditation. *Current Opinion in Psychology, 28,* 76-80.
22. Alda, M., Puebla-Guedea, M., Rodero, B., Demarzo, M., Montero-Marin, J., Roca, M., & Garcia-Campayo, J. (2016). Zen meditation, length of telomeres, and the role of experiential avoidance and compassion. *Mindfulness, 7,* 651-659.
23. Conklin, Q. A., King, B. G., Zanesco, A. P., Lin, J., Hamidi, A. B., Pokorny, J. J.,...Saron, C. D. (2018). Insight meditation and telomere biology: The effects of intensive retreat and the moderating role of personality. *Brain, Behavior, and Immunity, 70,* 233-245.
24. Mettananda, K. C. D., & Mettananda, S. (2024). Burden of disease scenarios for 204 countries and territories, 2022-2050: A forecasting analysis for the global burden of disease study 2021. *Lancet, 403*(10440), 2204-2256.
25. Black, D. S., & Slavich, G. M. (2016). Mindfulness meditation and the immune system: A systematic review of randomized controlled trials. *Annals of the New York Academy of Sciences, 1373*(1), 13-24.
26. Dispenza, J. (2019). *Becoming supernatural: How common people are doing the uncommon.* Hay House.
27. Akarsu, E., Korkmaz, H., Oguzkan Balci, S., Borazan, E., Korkmaz, S., & Tarakcioglu, M. (2016). Subcutaneous adipose tissue type II deiodinase gene expression reduced in obese individuals with metabolic syndrome. *Experimental and Clinical Endocrinology & Diabetes: Official Journal of the German Society of Endocrinology [and] German Diabetes Association, 124*(1), 11-15.
28. Lee, M. R., Schwandt, M. L., Bollinger, J. W., Dias, A. A., Oot, E. N., Goldman, D.,...Leggio, L. (2015). Effect of functionally significant deiodinase single nucleotide polymorphisms on drinking behavior in alcohol dependence: An exploratory investigation. *Alcoholism: Clinical and Experimental Research, 39*(9), 1665-1670.
29. Gałecka, E., Talarowska, M., Orzechowska, A., Górski, P., Bieńkiewicz, M., & Szemraj, J. (2015). Association of the DIO2 gene single nucleotide polymorphisms with recurrent depressive disorder. *Acta Biochimica Polonica, 62*(2), 297-302.
30. Park, E. J., Grabińska, K. A., Guan, Z., & Sessa, W. C. (2016). NgBR is essential for endothelial cell glycosylation and vascular development. *EMBO Reports, 17*(2), 167-177.
31. Kubota, S., & Takigawa, M. (2011). The role of CCN2 in cartilage and bone development. *Journal of Cell Communication and Signaling, 5*(3), 209-217.
32. Deutsch, D., Leiser, Y., Shay, B., Fermon, E., Taylor, A., Rosenfeld, E., ... Mao, Z. (2002). The human tuftelin gene and the expression of tuftelin in mineralizing and nonmineralizing tissues. *Connective Tissue Research, 43*(2-3), 425-434.
33. Lee, M. R., Schwandt, M. L., Bollinger, J. W., Dias, A. A., Oot, E. N., Goldman, D.,...Leggio, L. (2015). Effect of functionally significant deiodinase single nucleotide polymorphisms on

drinking behavior in alcohol dependence: An exploratory investigation. *Alcoholism: Clinical and Experimental Research, 39*(9), 1665-1670.

34. Zhang, H., & Song, J. (2021). Knockdown of lncRNA C5orf66-AS1 inhibits osteosarcoma cell proliferation and invasion via miR-149-5p upregulation. *Oncology Letters, 22*(5), 757.
35. Omary, M. B., Ku, N. O., Strnad, P., & Hanada, S. (2009). Toward unraveling the complexity of simple epithelial keratins in human disease. *Journal of Clinical Investigation, 119*(7), 1794-1805.
36. Aoki, J., Katoh, H., Mori, K., & Negishi, M. (2000). Rnd1, a novel rho family GTPase, induces the formation of neuritic processes in PC12 cells. *Biochemical and Biophysical Research Communications, 278*(3), 604-608.
37. Kaunitz, J. D. (2018). Magnetic resonance imaging: The nuclear option. Digestive Diseases and Sciences, 63, 1100-1101.
38. Leskowitz, E. (2014). Harvard doc to Wikipedia: You're not playing fair on alternative trauma therapy. *WBUR*. Retrieved from: https://www.wbur.org/news/2014/11/28/harvard-doc-to-wikipedia-youre-not-playing-fair-on-alternative-trauma-therapy
39. Wikipedia (2024a). *Acupuncture*. Retrieved 10/30/2024 from https://en.wikipedia.org/wiki/Acupuncture
40. Wikipedia (2024b). *Absurdly One Sided*. Retrieved 10/30/2024 from https://en.wikipedia.org/w/index.php?title=Talk:Acupuncture&oldid=1100285721
41. Ma, Y., Dong, M., Zhou, K., Mita, C., Liu, J., & Wayne, P. M. (2016). Publication trends in acupuncture research: A 20-year bibliometric analysis based on PubMed. *PloS One, 11*(12), e0168123.
42. Church, D., Feinstein, D., Palmer-Hoffman, J., Stein, P. K., & Tranguch, A. (2014). Empirically supported psychological treatments: The challenge of evaluating clinical innovations. *Journal of Nervous and Mental Disease, 202*(10), 699-709.
43. White, K. M., & Dudley-Brown, S. (2024). *Translation of evidence into nursing and healthcare*. Springer.
44. Church, D., Stapleton, P., Vasudevan, A., & O'Keefe, T. (2022). Clinical EFT as an evidence-based practice for the treatment of psychological and physiological conditions: A systematic review. *Frontiers in Psychology, 13,* 951451.
45. *Ibid.*
46. Church, D. (2014). Clinical EFT (Emotional Freedom Techniques) as single session therapy: Cases, research, indications, and cautions. In M. Hoyt & M. Talmon (Eds.), *Capture the moment: Single session therapy and walk-in service*. Crown House.
47. Ferriss, T. (2017). *Tribe of mentors: Short life advice from the best in the world*. Houghton Mifflin Harcourt.
48. Pressparty. (2013). *Katy Perry relaxes with meditation*. Retrieved from https://www.pressparty.com/pg/newsdesk/katyperry/view/86057/?isworld=y
49. Thomas-Mason, L. (2021). Watch David Lynch and Paul McCartney discuss the world of Transcendental Meditation. *Far Out Magazine*. Retrieved from https://faroutmagazine.co.uk/david-lynch-paul-mccartney-meditation-talk
50. Romm, C. (2015). Americans are more afraid of robots than death. *Atlantic*. Retrieved from https://www.theatlantic.com/technology/archive/2015/10/americans-are-more-afraid-of-robots-than-death/410929
51. Pinker, S. (2019). *Enlightenment now: The case for reason, science, humanism, and progress* (p. 157). Penguin.
52. United States Institute of Peace. (2013). *Peace events of the 20th and 21st centuries*. Retrieved from https://www.usip.org/sites/default/files/2017-01/Peace%20Events%20of%20the%2020th%20and%2021st%20Centuries.pdf
53. Yoder, K. (2022). St Joan of Arc: 15 quotes from her trial and interrogations. *Catholic News Agency*. Retrieved from https://www.catholicnewsagency.com/news/251395/st-joan-of-arc-15-quotes-from-her-trial-and-interrogations
54. Our World in Data. (2024). *Data Page: Homicide rates over the long term. Data adapted from Eisner (2014); WHO Mortality Database (2024)*. Retrieved from https://ourworldindata.org/grapher/homicide-rates-across-western-europe

55. Pinker, S. (2019). *Enlightenment now: The case for reason, science, humanism, and progress.* Penguin.
56. Eveleigh, R. (2023). The motley (star-studded) crew that paved the way for Norway's EV revolution. *Positive.News.* Retrieved from https://www.positive.news/society/the-crew-that-paved-the-way-for-norway-ev-revolution
57. United States Department of Agriculture (USDA). (2019). *Organic production.* Retrieved from https://www.ers.usda.gov/data-products/organic-production/documentation.aspx
58. International Energy Agency (IEA). (2024). *Carbon capture, utilisation and storage.* Retrieved from https://www.iea.org/energy-system/carbon-capture-utilisation-and-storage
59. NASA. (2023). *Aerosols: Small particles with big climate effects.* Retrieved from https://science.nasa.gov/science-research/earth-science/climate-science/aerosols-small-particles-with-big-climate-effects/
60. Economist. (2022). Adding bacteria can make concrete greener. *Economist.* Retrieved from https://www.economist.com/science-and-technology/2022/11/23/adding-bacteria-can-make-concrete-greener
61. Good News Network. (2019). *NASA happily reports the earth is greener, with more trees than 20 years ago–and it's thanks to China, India.* Retrieved from https://www.goodnewsnetwork.org/nasa-says-earth-is-greener-than-ever-thanks-to-china-and-india/
62. Corbley, A. (2024). *Norway's forests have more than tripled in a hundred years.* Retrieved from https://www.goodnewsnetwork.org/norways-forests-have-more-than-tripled-in-a-hundred-years/
63. Tabor, A. (2019). *Human activity in China and India dominates the greening of Earth, NASA study shows.* Retrieved from https://www.nasa.gov/centers-and-facilities/ames/human-activity-in-china-and-india-dominates-the-greening-of-earth-nasa-study-shows
64. NASA. (2024). *Leaf area index.* Retrieved from https://neo.gsfc.nasa.gov/view.php?datasetId=MOD15A2_M_LAI
65. Waidelich, P., Batibeniz, F., Rising, J., Kikstra, J. S., & Seneviratne, S. I. (2024). Climate damage projections beyond annual temperature. *Nature Climate Change,* 1-8.
66. Church, D., Vasudevan, A., & De Foe, A. (2024). The association of relational spirituality in an EcoMeditation course with flow, transcendent states, and professional productivity. *Advances in Mind-Body Medicine, 28*(3), 4-14.
67. Bodanis, D. (2021). *The art of fairness: The power of decency in a world turned mean.* Harry N. Abrams.
68. Economist. (2020). Why fair play pays. *Economist.* Retrieved from https://www.economist.com/business/2020/12/10/why-fair-play-pays
69. Wilks, M., McCurdy, J., & Bloom, P. (2024). Who gives? Characteristics of those who have taken the Giving What We Can pledge. *Journal of Personality, 92*(3), 753-763.
70. Church, D., & David, I. (2019). Borrowing benefits: Clinical EFT (emotional freedom techniques) as an immediate stress reduction skill in the workplace. *Psychology, 10*(7), 941-952.
71. Justsomething. (2020). *22 weirdest and dumbest US laws that are still in effect in 2020.* Retrieved from http://justsomething.co/the-22-most-ridiculous-us-laws-still-in-effect-today-2
72. Tanner, M. D. (2016). Too many laws, too much regulation. *Cato Institute.* Retrieved from https://www.cato.org/publications/commentary/too-many-laws-too-much-regulation
73. Church, D. (2015). *Psychological trauma: Healing its roots in brain, body, and memory.* Energy Psychology Press.
74. Church, D. (2009). The effect of EFT (Emotional Freedom Techniques) on athletic performance: A randomized controlled blind trial. *Open Sports Sciences, 2,* 94-99.
75. Chenoweth, E., & Stephan, M. J. (2011). *Why civil resistance works: The strategic logic of nonviolent conflict.* Columbia University Press.
76. Chenoweth, E. (2013). The success of nonviolent resistance campaigns. *Journal of Peace Research, 50*(3), 415-433.
77. Orme-Johnson, D. W., Cavanaugh, K. L., Dillbeck, M. C., & Goodman, R. S. (2022). Field-effects of consciousness: A seventeen-year study of the effects of group practice of the Transcendental Meditation and TM-Sidhi programs on reducing National Stress in the United States. *World Journal of Social Science, 9*(2).

Index